MATH FOR SECURITY

From Graphs and Geometry to Spatial Analysis

by Daniel Reilly

San Francisco

Printed in the United States of America

First printing

27 26 25 24 23   1 2 3 4 5

ISBN-13: 978-1-7185-0256-7 (print)
ISBN-13: 978-1-7185-0257-4 (ebook)

Publisher: William Pollock
Managing Editor: Jill Franklin
Production Manager: Sabrina Plomitallo-González
Production Editor: Miles Bond
Developmental Editor: Alex Freed
Cover Illustrator: Gina Redman
Interior Design: Octopod Studios
Technical Reviewer: Ricardo M. Czekster
Copyeditor: Rachel Monaghan
Compositor: Jeff Lytle, Happenstance Type-O-Rama
Proofreader: James M. Fraleigh
Indexer: BIM Creatives, LLC

For information on distribution, bulk sales, corporate sales, or translations, please contact No Starch Press® directly at info@nostarch.com or:

No Starch Press, Inc.
245 8th Street, San Francisco, CA 94103
phone: 1.415.863.9900
www.nostarch.com

Library of Congress Control Number: 2023002853

To my family. You are the reason for
all that I do.

About the Author

Daniel Reilly is a security researcher, analyst, and consultant based out of Seattle, Washington. He has worked in the security field for 20 years, more than half of which has been spent developing and managing operational security for small businesses.

About the Technical Reviewer

Ricardo M. Czekster is an assistant professor in computer science at Aston University in Birmingham, UK, where he focuses on dependability and cybersecurity with quantitative analysis. Having worked with security-driven tools and practical applications of modeling and simulation throughout his career, Ricardo is interested in modeling adversaries across systems, cyber threat intelligence, automated risk assessment, threat modeling, and intrusion detection. He has worked in research laboratories and organizations throughout the world, including Siemens in Princeton, the Laboratory for Foundations of Computer Science in Edinburgh, and the Secure and Resilient Systems in Newcastle upon Tyne.

Today's scientists have substituted mathematics for experiments, and they wander off through equation after equation, and eventually build a structure which has no relation to reality.

—Nikola Tesla

BRIEF CONTENTS

CONTENTS IN DETAIL

PART II: GRAPH THEORY AND COMPUTATIONAL GEOMETRY 25

3 SECURING NETWORKS WITH GRAPH THEORY 27

4 BUILDING A NETWORK TRAFFIC ANALYSIS TOOL 45

5 IDENTIFYING THREATS WITH SOCIAL NETWORK ANALYSIS 67

ACKNOWLEDGMENTS

A book like this can't exist without the effort of dozens of people, either directly or indirectly. When I decided to publish my work in 2016, I had no idea just how many people would pitch in to help me finish what I started. Then I started thinking about all the people I really owed my career growth to over the years. It has made me realize it would be impossible for me to list each of you individually, so let me start by saying a general thank-you to everyone who has contributed to any of my projects over the last 20 years. The hacker community has always been about the open sharing of knowledge. It's through this sharing and encouragement that I was able to develop the skills to be a security analyst. The organizers of conferences like DEF CON and local BSides events deserve a special thanks for all the effort they put in. Attending these events is a chance to bond with the members of our community and learn from one another. These events can also be chaotic whirlwinds behind the scenes, and organizing this chaos is no small feat! I'd also like to thank the Frank Lloyd Wright Foundation and Daniel Trujillo at the Artists Rights Society for helping me navigate the licensing process for the Guggenheim images used in the book. I've never worked copyrights, but you made the whole process exceedingly simple, which I appreciate!

I also want to say a huge thank-you to the team at No Starch Press. The time you've all put in to making my scribbles into a book is truly amazing. I never really understood the work it takes to produce a No Starch title, and now that I do, I appreciate my collection even more! Athabasca and Alex, you've both spent so much time revising my drafts that I can honestly say this book wouldn't exist without you! Jill, Miles, and Rachel, you've all taught me so much about the editorial process. I cannot think of a better team of people to have supported me through this process.

In my professional life there have been a lot of people who have helped me grow and evolve, but my teammates at Easy Metrics deserve special recognition. I seriously couldn't have accomplished any of this without all the knowledge and support you've all provided over the years we've known each other. Dean, Derek, Jessica, Josh, Owen, and Paul: you've each brought a special set of skills to the group dynamic, and I've learned so much from working with you over the years. I'm honored to count you all as colleagues and friends. Dan, you've been more than my manager; you've been my friend and sifu. I've learned more about business in our casual chats than most people pay a tuition for. With a team like ours, there's nothing we can't accomplish! Jay, what can I say other than thank you for everything, from teaching me how to run product development and encouraging me to take my security research more seriously, to being my sounding board for wild ideas and helping me rein them in just the right amount. You're a great mentor and friend, and there's always a spot open for you at my campfire.

On a personal note, there are a couple of people who have given so much of their time, energy, knowledge, love, and encouragement that I want to let them know how important they've been. Tony and Bam, my lawyer advised me that I cannot thank you for specific events that may implicate anyone in that thing that time, but you know. You're more than just my closest friends; you're the family that I needed when I needed it the most. Plus, I've learned more about security from hanging out with you two than any course, class, or book could ever hope to teach. In a lot of ways, this is all your fault!

Finally, the biggest thanks goes to my girlfriend and kids for putting up with me. I know I can be a lot to absorb at times. You've been subjected to impromptu math lectures and dragged to security conferences, or left at home when I traveled to one. You take care of literally everything while I get lost in weeks-long coding projects. You've done everything possible to support me through my career, and I love you all.

INTRODUCTION

Welcome to *Math for Security*, a book that is, by my estimation, unlike any other security book I've ever read. It isn't intended to introduce you to security topics like access control or encryption, nor is it going to help you prepare for the next certification exam. It will, however, strengthen your ability to examine the world around you and investigate security-relevant questions. The contents of the book sit at a crossroads between theoretical research, experimentation, and practical application. My goal is to introduce you to key mathematical fields through practical research topics. I find the easiest way to understand a complex theoretical principle is to see it in action.

I am not a mathematician by any stretch of the imagination. I am a security practitioner who loves reading theoretical research papers and has found, over the years, that there are not many resources for translating this theoretical research into a testable system, otherwise known as a *proof of concept*. I decided to write the book I wish I had when I started down this path. As you read it, you'll become familiar with the tools and procedures necessary to translate mathematical theory into security-relevant applications. You'll learn to accurately assess and communicate the limitations of your tools and procedures by examining the inevitable assumptions we must make when dealing with data in the wild. Most of all, though, I hope you'll learn to see mathematical theory in a new, totally practical light.

Who Should Read This Book?

Often, when I begin a discussion about applied mathematics in security, people say something along the lines of "but I don't work on encryption," which tells me many security practitioners misunderstand the role math plays in their day-to-day activities. The truth is, applied mathematics is at the heart of every modern security automation tool, not just cryptographic tools. My hope is that anyone with an interest in security and some Python programming experience will find something fun and informative in these projects. If you know enough Python to install and import packages, read and write files, and manage basic networking tasks, you should be set. You don't need a deep understanding of math to follow along, since we'll break down the formulas as we use them. If the title of the book caught your attention, chances are you're one of the people who should read this book!

What's in This Book?

I've organized the material into three parts. Part I: Environment and Conventions (Chapters 1 and 2) will help you set up a Python environment for coding and introduce some basic notation concepts. Part II: Graph Theory and Computational Geometry (Chapters 3–10) introduces the two main branches of math we'll be discussing. Finally, Part III: The Art Gallery Problem (Chapters 11–13) covers a larger project built to answer the classic art gallery problem, which will require both branches of math for our solution. Each chapter concludes with a "Summary" section that will cover other potential applications, background readings, and complementary math topics to help you continue on your own. Here's a brief overview of what to expect in each chapter:

Chapter 1: Setting Up the Environment We'll start by getting your environment set up using Anaconda or pip, depending on your preference and level of familiarity with Python. We'll discuss the importance of isolating development environments and cover how to access the Jupyter Notebooks included in the supplemental materials.

Chapter 2: Programming and Math Conventions Here we'll cover the special programming syntax and math notation you'll need to follow along with the examples. We'll cover Python's list and dictionary comprehensions as well as the `zip` and `unpack` functions. Then we'll dive into the math side of things with a refresher on Boolean algebra and set notation; we'll also look at the many different symbols that populate the math world.

Chapter 3: Securing Networks with Graph Theory This chapter will introduce the concepts around building and analyzing graphs using the NetworkX Python library. We'll discuss some of the basic types of graphs you're likely to encounter, as well as how to analyze some key statistics that will help you better understand the structure of your graphs.

Chapter 4: Building a Network Traffic Analysis Tool In this project, we'll build a graph to represent the communications across a computer network using the NetworkX and Scapy libraries. We'll then apply our understanding of graphs to uncover some interesting facts about computers on the network. We'll conclude with a proof-of-concept project in which you'll capture network packets and use them to create a graph of your own.

Chapter 5: Identifying Threats with Social Network Analysis This project focuses on analyzing people networks instead of computer networks. We'll build a graph using data simulating a social network like Mastodon. We'll discuss one of the ways connections form in social networks and dive into some practical research questions to identify interesting people. We'll conclude with a proof-of-concept project that will get you started collecting your own data for analyzing Mastodon timelines.

Chapter 6: Analyzing Social Networks to Prevent Security Incidents We continue our analysis of the social network from Chapter 5 by looking at what might happen in the future. We'll use randomness and probability to create a simulation of how messages might move through the network, who is likely to respond to whom, and on what topics. Finally, we'll cover our proof of concept, a two-player adversarial game to decide the fate of our social network.

Chapter 7: Using Geometry to Improve Security Practices In this chapter we switch to our second branch of math, computational geometry. We'll discuss how to represent different shapes using the Shapely library and common operations you'll encounter when working with shape data. The chapter will also introduce some physical security concepts, like resource planning and location, as we discuss representing our plans using geometry.

Chapter 8: Tracking People in Physical Space with Digital Information We continue our exploration of computational geometry by discussing its application in locating devices in the physical world. We'll cover the structure of network data and how to connect to the OpenCell API to gather geographic information about networks. We'll also discuss the ethics around device tracking. Our proof-of-concept project for this chapter will take in a sample set of tower data, solve for the overlapping coverage area, and return it as a bounded search area.

Chapter 9: Computational Geometry for Safety Resource Distribution In this chapter, we cover how to use Voronoi diagrams when doing resource analysis. We'll discuss the current distribution of fire stations around Portland, Oregon, and explore where a new station could have the most impact. We'll cover how to get the shape data for larger areas using the OpenStreetMap API. For the proof of concept, we'll create an application capable of programmatically recommending the location for a new fire station based on the current stations' responsibilities.

Chapter 10: Computational Geometry for Facial Recognition In the final project for computational geometry, we'll examine its application in facial recognition research. We'll cover what makes good images, how to process image data, different ways to measure interesting facial features with shapes, and how to find the key features for the best outcome. In this special two-part proof of concept, we'll develop a system to address both aspects of modern data science projects: model training and model application. We'll produce a system capable of processing an image set, train a facial classifier, and ultimately apply our method to three faces to see if we can properly identify them using only computational geometry.

Chapter 11: Distributing Security Resources to Guard a Space This chapter begins our exploration of the art gallery problem and the requirements of a larger project. We'll cover the theory and existing research that underpins our solution. Then we'll discuss how we can combine graph theory and computational geometry to improve upon the basics for a more realistic answer. The code in this chapter will serve as the base for the rest of the project and covers generating the actual solution we're interested in.

Chapter 12: The Minimum Viable Product Approach to Security Software Development Here we'll expand our project from Chapter 11 beyond a simple proof of concept by adding more advanced features that will improve the user experience. We'll discuss speeding up our application with parallel programming. We'll touch on using PyGame to handle graphics and user interactions. Finally, we'll go over the example application that comes with the book and explore how you can use it as a starting point for your own art gallery problem solver.

Chapter 13: Delivering Python Applications We'll conclude our art gallery problem project with a discussion of modern software delivery methods. This is a large topic, so I've picked a few that I think everyone should be familiar with, from packaging your application as a library to delivering it as a cloud service. We'll discuss some of the pros and cons of each option as well as its impact on your ability to monetize your application.

Why Python?

Python checks all the boxes for what we'll be covering in this book. The Python language has a long history of success in the security community and there's already a plethora of tools and books that use it. Operational flexibility has been cited as one of the primary reasons for Python's heavy adoption in the security community. Python performs well across a wide variety of platforms, from tiny single-board computers (like Raspberry Pi) to massive computing clusters and just about everything in between. With Python, you'll be able to integrate your ideas with a lot of existing work.

Python also has strong ties to the applied mathematics communities. In today's computer-centric world, applied mathematics is usually expressed using one or more high-level programming languages. Of all modern programming languages, Python has emerged as a leader in mathematical and scientific programming because it's easy to learn but almost unlimited in its expressive power. Python 3 is a natural choice for the exploratory research we'll be doing because it has many libraries and features that will help us implement the various algorithms required to perform in-depth analysis.

That isn't to say that Python doesn't have its shortcomings. Python is an interpreted language, which means a program (called the *interpreter*) sits between the code you write and the system executing that code. The interpreter's job is to translate your instructions into commands the underlying system understands.

Having the interpreter between what you write and what your system receives introduces a few problems. First, the additional layer between the instruction and the execution adds processing time and memory overhead. Second, the interpreter is itself a single application running in a process on a core of your machine. Your code is then executed inside the context of the interpreter's process, which means your entire application is restricted to the tiny fraction of the system that the OS allocates to a process, even if you have an octocore beast with enough RAM to store the congressional library. There are programming tricks you can employ to sidestep limitations (such as distributed processing, which we'll discuss in Chapter 12), but ultimately Python will never be as fast and free as a compiled language like C. The dirty truth is that a lot of the computationally heavy functions in data science libraries like scikit-learn and NumPy are wrappers for compiled C programs under the hood.

Shortcomings aside, Python is nevertheless the best choice for our goals. I'll introduce some common idioms in Chapter 2 that we'll use throughout the rest of the book, and I'll explain the code as necessary in the following chapters.

Information Accessibility

Throughout the book, I'll do my best to represent the data included in figures in multiple distinct ways, such as numbers and symbols combined with color gradients. I chose to add the shape or number representation along with the color to help convey the point more clearly, since the images in the printed text are grayscale. The choice to use another indicator along with colors is also one of accessibility. Having to write for a book using only grayscale images has helped me realize how much information we try to convey with color and how unfair that is to people who cannot distinguish a large range of colors. Whenever we create a data visualization, we should strive to give people a number of ways to distinguish the key points, and not over-rely on pretty color gradients to do the talking for us. If you have accessibility needs or concerns for the information presented here, please contact me on the book's GitHub (*https://github.com/dreilly369/AppliedMathForSecurityBook*), and I'll do my best to provide a solution.

Online Resources

The supplemental materials and accompanying Jupyter Notebooks for this book are available at *https://github.com/dreilly369/AppliedMathForSecurityBook*.

PART I

ENVIRONMENT AND CONVENTIONS

1

SETTING UP THE ENVIRONMENT

Let's start by setting up the programming environment we'll use throughout the rest of the book. Python is flexible enough to run across a number of platforms, so I can't cover all possible installation and configuration options. That said, because some of the problems we'll be analyzing can be computationally expensive, I'll assume you're using a laptop or desktop to do your experimentation, and not a tablet or phone. A multicore CPU will help speed up some of the processing. While it's not necessary (and I won't be using them), some libraries can also leverage modern GPUs, so I encourage you to experiment with them. Finally, several operations can be memory intensive. I recommend having at least 4GB of RAM available, but 8GB or more is preferred. With every environment, you have to balance implementation cost against the time it takes to generate solutions. In Chapter 13 we'll examine distributing the problem across many smaller platforms.

I'll present two setups: one simple, one advanced. If you're new to Python programming, I recommend you start out with the simple setup, which uses Anaconda for package and environment management and installs an integrated development environment (IDE) named Spyder. The Spyder IDE is specifically tooled toward mathematical and scientific applications, making it an excellent choice for the upcoming projects.

If you're familiar with the nuances of environment and package management and already have a Python environment configured, the advanced setup covers using a virtual environment to isolate your experimental workspace from the rest of your production tools, as well as manually installing the required packages.

Simple Environment Configuration with Anaconda

We'll start by installing Anaconda, a platform designed for managing multiple Python environments painlessly, even for people who don't have a background in system administration. Anaconda will make it easy to install the packages we need, update them over time, and ensure environment dependencies remain consistent. Anaconda has been specifically designed for data science and machine learning workflows. There are installers available for Linux, Windows, and macOS. Head to the Anaconda distribution page (*https://www.anaconda.com/distribution*) and download the most recent installer version for your platform.

Now let's walk through the instructions for Linux, Windows, and macOS.

Linux

The download you receive from the Anaconda link is actually a shell script to assist in downloading and configuring the necessary packages. You should *not* run this script as a privileged user (like `su` on Debian, for instance). Start by opening a terminal in the directory where the install script is located. You can execute the installer with the following commands:

```
$ chmod +x Anaconda3-202X.0X-Linux-x86_64.sh;
$ ./Anaconda3-202X.0X-Linux-x86_64.sh;
```

To begin the installation, you need to mark this script as executable using `chmod +x`. Make sure to change the name of the script to match the version you downloaded. You can then run the installer using the default shell interpreter. During setup you'll be asked to confirm a few installation options. In most cases, the default options are good enough. If you plan to change any of the defaults, take the time to read the documentation—some options can have unforeseen consequences. After the installation completes, you can verify everything is working using the newly installed conda utility. Open a new terminal and issue this command:

```
$ conda info
```

You should see output similar to the following:

```
     active environment : base
    active env location : /home/dreilly/anaconda3
            shell level : 1
       user config file : /home/dreilly/.condarc
 populated config files :
          conda version : 23.X
    conda-build version : not installed
         python version : 3.X
       virtual packages : __glibc=2.23
       base environment : /home/dreilly/anaconda3  (writable)
           channel URLs : https://repo.anaconda.com/pkgs/main/linux-64
                          https://repo.anaconda.com/pkgs/main/noarch
          package cache : /home/dreilly/anaconda3/pkgs
                          /home/dreilly/.conda/pkgs
       envs directories : /home/dreilly/anaconda3/envs
                          /home/dreilly/.conda/envs
               platform : linux-64
             user-agent : <platform-user-agent-string>
                UID:GID : 1000:1000
             netrc file : None
           offline mode : False
```

There's some good information here. First, there's `user config file`, the location of the user configuration file. Editing this file will allow you to personalize Anaconda; it's worth looking into if you plan to do a lot of work in it. The next two pieces are `conda version` and `python version`. Compare `conda version` against the latest available Anaconda version to ensure you have the most up-to-date tools possible. The `python version` is the default Python interpreter that Anaconda will use when creating environments. You can set specific Python versions for each environment, but ensuring your default is set to your preferred version can save you some time if you forget to specify a version when you create an environment.

The `channel URLs` field tells you the remote locations conda will check when trying to install new packages. Be careful when modifying this list. Adding untrusted source repositories poses a security risk if an attacker replaces a legitimate package like pandas with a malicious version. It's also a good idea to audit this field from time to time to make sure it doesn't include any unrecognized channels. The `package cache` field shows you where Anaconda will store package information for the libraries it has installed. Since multiple environments will likely request the same version of packages, Anaconda builds a cache of the known packages to speed up future installations and reduce creation times for similar environments. The final piece to take note of is the `envs directories` field, which tells you where on the system Anaconda will store the files related to defining each environment, including a copy of the package versions installed for each. Knowing where this information is stored will come in handy if you have to troubleshoot a package conflict in a specific environment (although conda has tools to help with this as well).

At this point, your base environment is set up and you're ready to configure your research environment. You can skip to the "Setting Up a Virtual Environment" section later in the chapter.

Windows

When you run the Anaconda install script for a Windows machine, you'll be greeted with a typical Windows-style install prompt, similar to Figure 1-1.

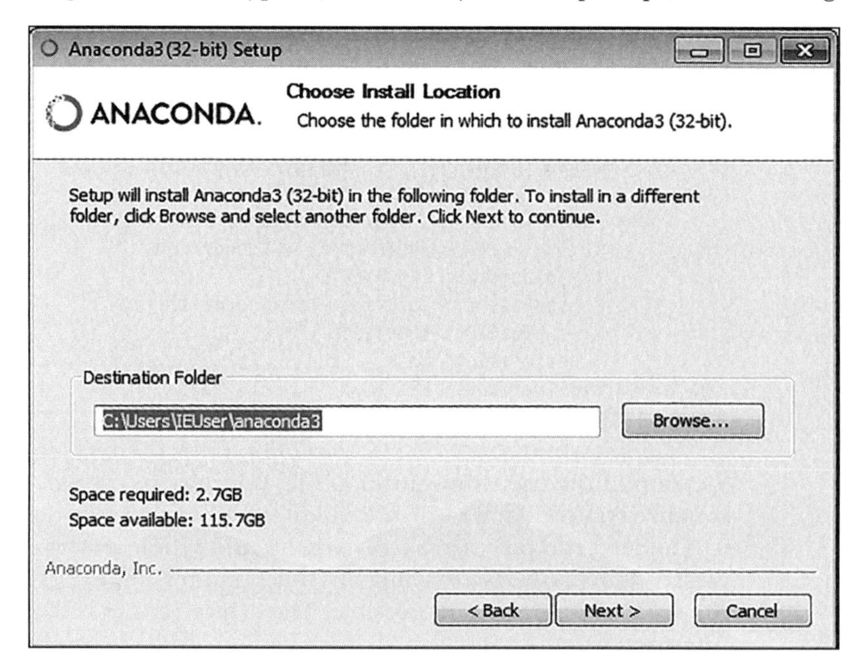

Figure 1-1: The Anaconda installer on Windows

Select the directory where you'd like the base application to reside. If your system has a large secondary drive and a small primary solid-state drive (SSD), make sure you install Anaconda on the larger drive. With multiple environments and interpreters and package versions, it can grow quite big over time.

The rest of the installer will walk you through configuring Anaconda. The defaults are typically fine for most cases. If you plan to change any of the defaults, take the time to read the documentation—some options can have unforeseen consequences. You may need to give the installer permission to make changes (via the User Account Control pop-up). In some instances you may get an error message when the GUI attempts to start. You can often fix this issue by explicitly telling conda where to find the proper executable. Open a run prompt (you can use the key combination WIN-R on most keyboards), type **`cmd.exe`**, and then press ENTER. Navigate to the *scripts* subdirectory inside the directory where you installed Anaconda. If you used the install directory in Figure 1-1, for example, the path to the scripts directory would be *C:\Users\IEUser\anaconda3\scripts*. Move into that folder like this:

```
$ cd C:\Users\IEUser\anaconda3\scripts
```

Then run the following commands:

```
$ activate root
$ conda update -n root conda
```

The first command tells Anaconda to activate the root environment, which is created during installation and contains the default Python version and some basic packages. The second command tells Anaconda to update the root environment's version of the conda application. You'll be asked to confirm the update by pressing Y, and then Anaconda will install the latest version of conda. Finally, you can update all the packages in the root environment using the updated version of conda like so:

```
$ conda update --all
```

Once again, you'll be asked to confirm the update by pressing Y. After the setup and updates have completed, you can access an Anaconda Navigator interface similar to the one in Figure 1-2.

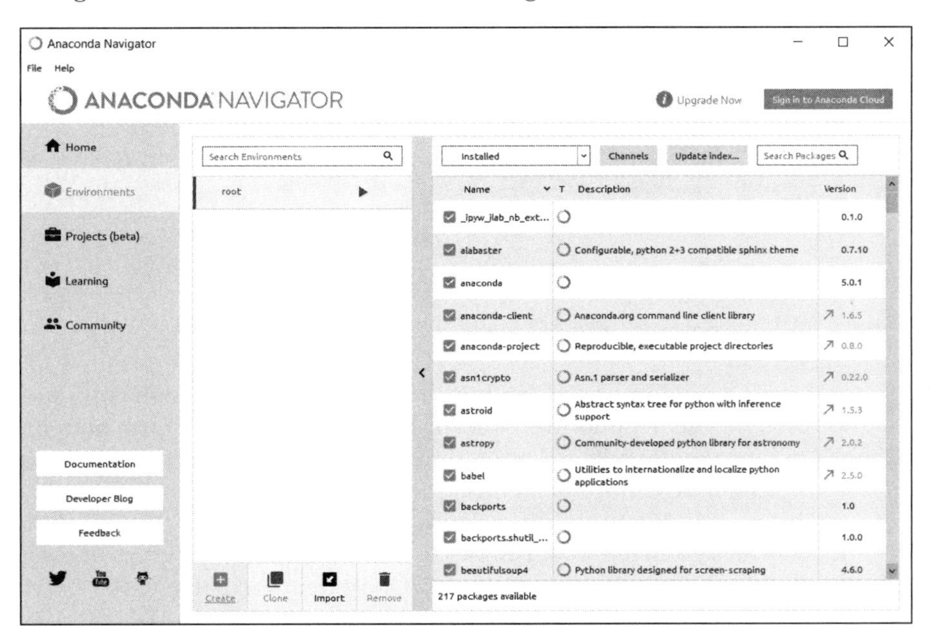

Figure 1-2: The Anaconda Navigator interface

The Navigator interface can be used for package management, virtual environment management, and more. It's the command center for administering your Anaconda installation. Even though Windows installations do have access to the conda utility (it's what the GUI relies on under the hood, after all), most never need to use it directly since all the most useful tools have been wrapped in a pleasant interface. One benefit of the Windows install is how easy it is to create virtual environments. Simply click the **Environments** tab and then click **Create**. Enter a name for your new environment and pick the appropriate interpreter. That's it! Now you can skip to the "Setting Up a Virtual Environment" section.

macOS

When you run the *.pkg* file downloaded from the Anaconda link, you should see something like Figure 1-3.

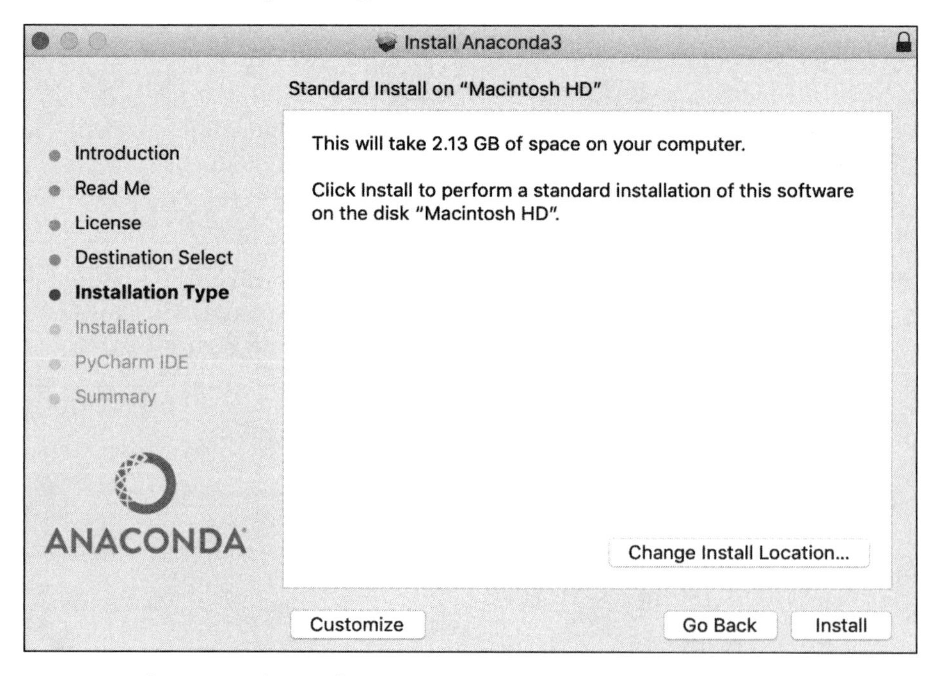

Figure 1-3: The Anaconda installation screen

The rest of the installer will walk you through configuring Anaconda. The defaults are typically fine for most cases. If you plan to change any of the defaults, take the time to read the documentation—some options can have unforeseen consequences.

Note that the recommended IDE (Spyder, covered shortly) isn't natively supported under macOS. It's available using the port package manager, though. If you're on a macOS system, you can elect to skip the Spyder setup described in this chapter and choose a Python IDE specifically designed for use with your system. None of Spyder's advanced features are necessary to view or run the code examples, so any modern IDE should suffice.

Setting Up a Virtual Environment

Anyone who has worked in Python for long enough has seen the inevitable chaos that comes from working on too many projects in one space. Different packages will require different versions of the same dependency. One project requires a different interpreter than another. It can be a real mess, and one you typically won't notice until it's too late and you're spending the weekend resolving your poor choices.

To avoid all the headache and lost time, you can simply break your projects up into separate virtual environments. You can think of a virtual environment as an isolated Python world. Packages in one virtual environment are blissfully unaware of packages in other virtual environments. Interpreters are automatically changed to the proper one for the environment. Life is once again harmonious. In your preferred terminal, enter the following command to have conda create a new environment for your projects.

```
$ conda create -n env_name python=3.X anaconda
```

Here `env_name` is the name you want to give your environment; change this to suit your preference. You can specify a Python version using the `python=version` syntax. The example will configure a Python 3.*X* interpreter as the default for your new environment. Press Y to proceed. Anaconda will make a copy of the Python interpreter and all the associated libraries to the *environments* subdirectory. The exact path to this directory will depend on your OS and where you installed Anaconda. On my system, the path to this environment becomes */home/dreilly/anaconda3/envs/researchenv/*.

Take a moment now to locate this directory and make a note of the full path to the folder. You'll need this in a few minutes to properly configure the IDE. To activate your new virtual environment and begin working within it, issue the following command in your terminal:

```
$ conda activate env_name
```

When you activate an Anaconda virtual environment, the management application modifies your operating system's underlying environment variables to accommodate the separation between projects. The `PATH`, `PYTHONPATH`, and other variables are updated to point at the specific isolated Python setup you've created. Your terminal prompt should change to tell you which virtual environment you're currently working in. You may also want to verify the setup by checking the `conda info` command.

```
$ conda info -e
```

The result should be a list of all the virtual environments you currently have defined with an asterisk next to the currently active environment.

Installing the IDE with Anaconda

Spyder (*https://www.spyder-ide.org*) is a scientific and mathematics programming environment written in Python, for Python. It's been designed by a dedicated group of programmers, scientists, engineers, and data analysts to meet the demands of their work. The best part about the Spyder IDE is the fact that it comes available through Anaconda. You can install it from your terminal with the following command:

```
$ conda install -c anaconda spyder
```

Once the install completes, you can run the IDE from the console with the command `spyder`.

Now that you've set up the virtual and coding environments, you're almost ready! You can give yourself two cool points and skip to the "Jupyter Notebooks" section of this chapter.

Advanced Setup

Using the following setup, you can isolate your experimental work from your production workspace using virtual environments. It assumes you've already installed Python 3 correctly on your system and you're familiar with using the pip utility to install packages. A *setup.py* script is provided in this book's GitHub repository for your convenience.

Setting Up virtualenv

The virtualenv is a Python package manager for isolating different projects' dependencies. Using virtualenv allows you to avoid installing Python packages globally, which could break system tools or lead to inconsistent dependent libraries in other projects. You'll be installing virtualenv using the pip utility. Enter the following command in your terminal:

```
$ python -m pip install --user virtualenv
```

Depending on your install, on Windows machines, you may need to change the command to the following:

```
$ py -m pip install --user virtualenv
```

After installing the virtualenv module, you can create the environment. The details of this process have changed over the years, but the idea remains the same. Since Python 3.9 this is the preferred method to create virtual environments:

```
$ python -m venv path_to_new_environment
```

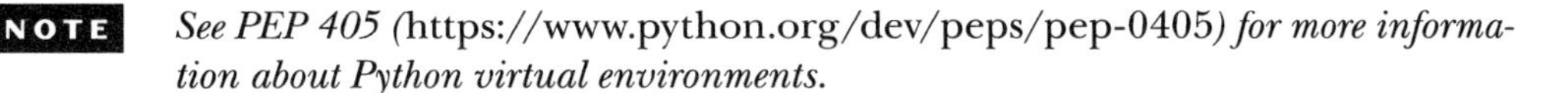
NOTE *See PEP 405 (*https://www.python.org/dev/peps/pep-0405*) for more information about Python virtual environments.*

Installing the IDE without Anaconda

You can install the Spyder IDE in many different ways depending on the platform you operate on. Debian Linux users may be interested in installing via apt-get instead of Anaconda. Spyder's official Debian package is available on the Debian package repository:

```
$ sudo apt-get install spyder3
```

This will install the spyder3 application to */usr/bin/anaconda3*. You can verify this location with the following command:

```
$ which spyder3
```

It's possible to install Spyder using pip, but it's not recommended (it can get quite complex). See the Spyder installation guide (*https://docs.spyder-ide.org/installation.html*) for more information.

If you're on Windows you can get the Spyder IDE, along with a suite of other useful libraries, within the Anaconda Navigator. Under the Home tab you'll see a sample of applications that integrate with Anaconda. Click **Install** under the Spyder application to begin the process. Once the installation is complete, the Install button will be replaced with a Launch button that will start the IDE.

WinPython (*https://winpython.github.io*) is another scientific Python distribution that is similar to Anaconda in that it includes all the most sought-after scientific packages and utility libraries. One notable difference, though, is the lack of a package management tool like Anaconda's conda utility. It's this lack of beginner-friendly tooling that ultimately made me decide to use Anaconda for this book. If you're already using WinPython you'll be able to follow along with the projects, but if you're starting a brand-new environment I highly recommend going the Anaconda route.

Jupyter Notebooks

Congratulations—at this point your development environment is complete! There's one more tool, however, that you can optionally install to get the most out of this book: the Jupyter notebook server. Several of the projects in this book are accompanied by interactive Jupyter notebooks that include more detail about the mathematical formulas, discussions about creating the graphics used in the text, and pieces of template code you can use to speed up your own application development. If you followed the Anaconda setup on Debian Linux, I have good news: you already have the Jupyter notebook server installed. If you installed Anaconda under Windows, you can use the Anaconda Navigator to get the application. Navigate to the **Home** tab and click **Install** under the Jupyter Notebook app. Once the Install button changes to a Launch button, you're ready.

You can also install Jupyter in your virtual environment manually with the following pip commands:

```
$ pip install --upgrade pip
$ pip install jupyter
```

You can then start a notebook server from the terminal using the following command:

```
$ jupyter notebook
```

Or on Windows you can simply start the notebook server using the Anaconda Navigator GUI as I mentioned before. Either way, Jupyter will open a web browser that displays the contents of the directory where you ran the command. You can use the web interface to create new notebooks or open existing ones.

If you're new to the concept of interactive notebooks, I urge you to take this opportunity to learn what they're all about. Jupyter is a web application that allows you to create and share documents that contain live code, visualizations, and text formatted in HTML or the more compact Markdown syntax. By combining Markdown, code, and output, we can automatically generate beautiful reports, informative analysis, and working proofs of concept all in one place. Code snippets can be run independently, but data persists between code in different blocks (called *cells*).

I regularly use notebooks while I'm writing and testing individual functions for a project because I can configure the variables once and then repeatedly test the function without needing to rerun any other code. As you'll see, I also used them extensively when writing these chapters to generate most of the figures and code blocks. Being able to run certain pieces of code independently came in handy when I was tweaking and rendering the figures.

Summary

Planning and building a solid Python environment may not be the most exciting reading topic in the world, but making sure all your tools are in place and ready to work will make jumping into the rest of the book much simpler. Whether you chose to go the simple route with Anaconda or the more advanced route with virtualenv, you should now have an isolated area to follow along with the code in the upcoming projects. In the next chapter, we'll finish our last bit of housekeeping by clearing up some of the programmatic and mathematical syntax used throughout the rest of the book. If you're already very comfortable with Python and know your sigmas from your deltas, feel free to skip ahead to Chapter 3.

2

PROGRAMMING AND MATH CONVENTIONS

Now that you have an environment to work in, let's discuss the programming language we'll be using. This book assumes you have a basic familiarity with programming concepts such as loops, variables, conditionals, and functions, so this chapter isn't meant to be a comprehensive introduction to Python. Rather, it's intended to illustrate some finer points that will help you understand the examples in this book and in other tutorials. Of course, the programming in the book centers on implementing math concepts from research papers, so it's equally important that we have a common understanding of the notation used in that type of material.

Syntactical Constructs

Although this book isn't meant to serve as an introduction to Python programming, there are some useful syntactical constructs you should be familiar with before diving in. It's important to balance the use of these more advanced features to maintain the readability and understandability of your code. I've used the constructs described in this section throughout the book's projects for code brevity. If you're still learning Python, some of the syntax may seem daunting at first, but once you apply these constructs a few times in your own code you'll wonder how you ever got anything done without them!

List Comprehensions

Comprehension constructs come in handy whenever you need to create a list (or dictionary) of values by iterating over some code. The simplest use case is to apply a function to every element in a list. For example, suppose you have a list of strings, called `names`, that you want to convert to uppercase. You could use a loop similar to the following:

```
names = ["bob","mike","tom","mary"]
names_2 = []
for n in names:
    names_2.append(n.upper())
```

But there are a couple of downsides to this method. First, it requires two lists: one for inputs (`names`) and one for outputs (`names_2`). When you have very large lists of complex objects, copying the whole list in memory for a simple change like this is inefficient. You could add more code to overwrite names with the contents of `names_2` and then free up the memory used by `name_2` explicitly, but that's a pain and makes the code messy. This leads us to the second problem: the code takes up more lines than necessary. In complex applications you may have hundreds of functions to maintain. Keeping your code concise saves you time when writing and modifying your code in large code bases. You could condense the whole process into a single line like so:

```
names = [n.upper() for n in names]
```

In a list comprehension, the first variable defines what will be stored in the resulting list. In this case `n` is a string we're converting to uppercase, hence `n.upper`. We define the values for `n` using the `for` statement to its right, which iterates through the `names` list and selects each `n`. The resulting list is then assigned directly to the `names` variable, overwriting its previous value and freeing up the memory used by the loop automatically. We can add some conditional logic to the end of the statement to filter the results. For example, suppose we want to also filter the `names` list on whether the first letter of each string is `m`. Once again, you could write this as a loop with a block of code like the following:

```
m_names_upper = []
names = ["bob","mike","tom","mary"]
for n in names:
    if n.startswith("m"):
        m_names_upper.append(n.upper())
```

Now within each iteration of the for loop we have a conditional statement to check if the string starts with m and, if so, it appends the uppercase version to the m_names_upper list. This method suffers from the same drawbacks as the previous loop example, though, and takes up even more space! This code can also be shortened to a single line with a list comprehension:

```
m_names_upper = [n.upper() for n in names if n.startswith("m")]
```

We've added some conditional logic to the end of the list comprehension statement to filter the results; again, we'll add n.upper only if n starts with an m.

The two examples will produce the same output; it's open for debate which is simpler or easier to explain, so the approach you choose will often be a matter of preference. Note that you can add an else condition to the list comprehension to control what happens in the event the if returns False. In cases where the conditional logic is False—for example, when a function expects two lists to be the same length so it can do some pairwise manipulation on the values—you might want to add a static value to your output list rather than excluding the element altogether. Let's say we want to rename everyone whose name doesn't start with m to marcus. With a traditional loop structure, this requires adding an else block after the if block to handle inserting the value. I'll skip that, as we've all seen else blocks in Python (hopefully), and we're discussing list comprehensions. Adding the additional else clause to a list comprehension changes the syntax so the if...else logic comes before the for loop logic, which looks like this:

```
m_names = [n if n.startswith("m") else "marcus" for n in names]
```

As you can see, the if statement moves directly after the variable to be stored, n. You then add the else condition, which defines the value to be added to the output list m_names in cases where the if statement is False. In this example, if a name starts with an m, it will be added to the output list; otherwise, the string literal marcus will be added instead.

There are plenty of more practical uses for list comprehensions, and you'll see them sprinkled through the code in this book and other examples, so it's best to become familiar with them and how they translate into more traditional looping code structures. It's also important to understand the limitations of list comprehensions. If your conditional selection logic is fairly complex, you might want to consider abstracting it to its own function. You can then use the function in a list comprehension to apply it to each element (a method you'll see used in this book as well). The only real limitations within the list comprehension itself are that you can apply only one conditional statement, and the output at the end of each iteration must be something that can be included in a Python list object.

Dictionary Comprehensions

Dictionaries are one of the most prevalent data structures in Python. They are used for everything from simple key/value pairing up to handling complex, user-defined classes. Luckily they are also extremely easy to work with, and a lot of the concepts we just discussed for list comprehensions apply equally well to defining dictionary objects, though you'll tend to run across fewer examples in the wild. As an illustration, consider the following code, which merges a list of keys with an accompanying list of values:

```
keys = ['Red','Blue','Green','Yellow']
values = [1,2,3,4]
out = {}
for i in range(len(keys)):
    out[keys[i]] = values[i]
```

Once again, the power of dictionary comprehension allows us to shorten this code block to a single line:

```
out = {keys[i]: values[i] for i in range(len(keys))}
```

You may notice there are only a few differences between the list comprehension and the dictionary comprehensions syntax. Python's developers have done this to make them both easier to understand in relation to each other. If you understand list comprehensions, then you understand 90 percent of dictionary comprehensions. The two things to remember are: dictionaries are defined using braces {} instead of square brackets, and the left part before the `for` keyword expresses both a key and a value, separated by a colon. Here, we define our key as the *i*th value in the list called `keys`. We're assigning the value for that key from the second list, `values`, using the same index, *i*. We then define a `for` statement that loops over the integer values between 0 and the length of the `keys` list using the `range` function. This is an example of a function that naively assumes both the `keys` and `values` lists are the same length. If `keys` is ever longer than `values`, this code will raise an error when it attempts to access a nonexistent index on `values`.

You can also use a function to define both the key variable and the value variable, like so:

```
out = {keys[i].upper(): float(values[i]) for i in range(len(keys))}
```

In this example we've changed the previous code so all the keys are converted to uppercase and all the values are cast to `float`. We could, of course, replace these functions with anything that suits our needs. As long as the result for each function can be used as a key or value in a regular Python dictionary, the function can be used. The same caveat applies to this conditional logic as with list comprehensions. If you need more complex logic, you should consider abstracting the key and value definition into its own function, like so:

```
out = {k: v for k,v in my_logic(keys)}
```

Here the keys and values are being defined by the return value from the function call `my_logic(keys)`. It doesn't matter what this function does; all that matters is that the function returns a list that contains tuples (or nested lists) that each contain two items. The first item in each tuple will be treated as the key `k`, and the second item as the corresponding value. And there's the dirty little secret of dictionary comprehensions: they're really just list comprehensions in disguise! You may have already realized this because the `range` function produces a list as well. This is a handy shortcut when you want to run a function on every item in a list and store the result as a dictionary that relates the item with the result of calling the function on that item.

Zipping and Unpacking

Python's built-in `zip` function returns an iterator of tuples, where the *i*th tuple contains the *i*th element from each argument passed in. Suppose you wanted to combine the previous list of names stored in `names` with the list of colors and numbers stored in `keys` and `values`, respectively. The `zip` function will allow you to efficiently generate these combinations as a list of tuples, again with a single line of code:

```
a = zip(names, keys, values)
print(list(a))
```

The variable `a` now holds a `zip` object that, when cast to a list, will contain triples like `("bob","Red",1)`. It's important to pay attention to the order in which you pass the lists, as left-to-right processing is guaranteed. Also be aware that the iterator stops when the shortest-length input is exhausted. Unpacking is the reverse of the `zip` function, but it's a behavior of Python rather than a function that you need to call. For example, you can unpack the three values that make up the first item in the `zip` object `a` like so:

```
person, key, value = list(a)[0]
```

This line of code takes the three values in the first tuple in `a` and assigns them to the three variables (left-to-right order) on the left side of the equal sign. You must have the same number of variables as items in the tuple, or the code will raise an exception. Zipping and unpacking are useful when you want to transport data around your application. Rather than defining multiple variables to hold the three output lists, we can apply the `zip` function to return a single `zip` object, which preserves the relationship between all three lists. You can then iterate over the results and unpack the values into distinct variables as needed.

The rest of the code you see will be pretty standard Python. I'll point out the syntax for specific libraries as we come to use them in our projects.

Next, we'll dive into the exciting world of math notation. As you'll see, understanding the symbols used plays a crucial role in advanced mathematics.

It provides a flexible form of shorthand that makes the formulas easier to remember. Unfortunately, as with programming, these symbols can take on multiple meanings. Mathematical notations often have a dual nature since they can signify both the math being performed and variables on which it's being performed.

Mathematical Notation

Mathematical notation is a tricky beast, especially when you intend to cover more than one field in a text. The reason is that a lot of mathematical symbols are *overloaded*, meaning they have many possible meanings, and to know which meaning applies, you need to look at the context. A perfect example of this is the Greek letter theta, θ. In machine learning literature, θ usually refers to the set of feature weights calculated for a set of data. This "hypothesis" function is commonly seen in the first weeks of machine learning, when linear regression models are covered.

$$h_{(x)} = \sum_{i=0}^{n}(\theta_{(i)}\, X_{(i)})$$

However, in geometry and trigonometry, θ is often used as a "degrees" variable—for example, in the Pythagorean theorem, as shown in Figure 2-1.

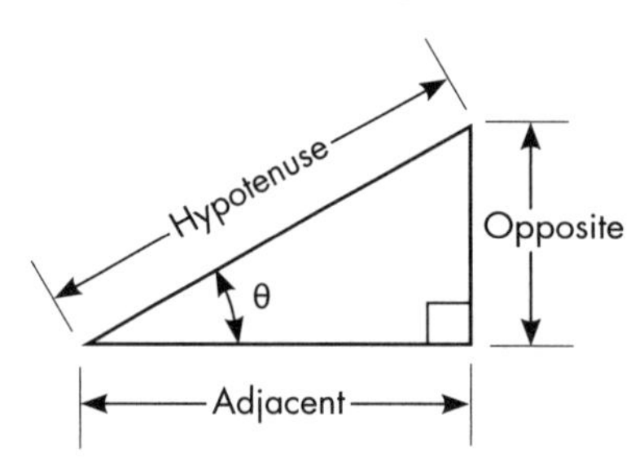

Figure 2-1: The Pythagorean theorem using theta to represent an angle in degrees

Where possible, I stick to the notation used in the prevailing material on the subject. In cases where the notation becomes overloaded, I've included explanations of the intended meanings.

Boolean Notation

One of the most often applied (and most often overlooked) concepts you'll see in applied mathematics is Boolean algebra. A *Boolean* is a primitive data type that can hold only one of two possible values at any given moment. The value is determined by the *logical statement* that accompanies it. For example, we can ask if two numbers, *x* and *y*, are equal to each other. The answer to this question will always be yes or no, depending on the inputs. There will never be a case that produces a third answer.

Boolean symbols are most often denoted in the form of a *truth table*. In the interest of saving space, I haven't included any full truth tables in this book; you can find many excellent references for these tables online. We'll focus instead on the symbols used to write the logical statements and how you can interpret them. Table 2-1 covers each of the major Boolean algebra symbols and the intuitive meaning, with an example of the logic applied to a statement.

Table 2-1: Boolean Logic Examples

Notation	Meaning	Example
AND: $A \wedge B$	Both *A* and *B* must be `True` for the statement to be `True`. Algebraically, $A \times B$ or just *AB*.	*Cats are mammals* AND *planes fly* is `True` since both statements are `True`. If either statement is `False`, the whole statement is `False` (equal to `0`).
OR: $A \vee B$	If either *A* or *B* or both *A* and *B* are `True`, this statement is `True`. Algebraically, $A+B$.	*Cats are dogs* OR *birds are cats* is `False` because neither *A* nor *B* is `True`. Changing either or both to `True` statements would change the result.
NOT: $\neg A$, $\grave{A}$	This statement is `True` when the statement *A* is `False`. It is an inverter that outputs the opposite of its input.	If *A* represents the statement *cats are fish*, this could be written as *NOT cats are fish*. *NOT cats are fish* is `True` because the statement *cats are fish* is `False`.
XOR: $A \otimes B$	If either *A* or *B* but NOT both *A* and *B* are `True`, this statement is `True`. Algebraically, $(A\grave{B})+(\grave{A}B)$.	*The shirt is red* XOR *the shirt is blue* is `True` if the shirt is either red or blue, but not both red and blue.

These deceptively simple components can be used to describe the most arbitrarily complex systems known to humankind through the use of chained Boolean statements. It's very important to be aware of the order of operations and placement of parentheses in Boolean expressions. Operator precedence is the same as in traditional algebra, with fewer functions. Any operations inside parentheses are handled first, followed by any AND conditions (multiplication), and finally, OR statements (addition). Negations are handled during the final step of the operation being negated, so they don't need any special handling. Traditionally we evaluate multiple parentheses from left to right.

Most runtime environments, like Python's interpreter, stop evaluating as soon as the truth of a statement can be determined. They won't continue to check additional statements that can no longer change the result. For example, the logical statement

$$(\textit{Cats fly} \wedge \textit{dogs bark})$$

will first check the *Cats fly* statement. Since this is `False`, the value of the second statement can't impact the result of the AND statement. Python determines that the total statement is `False` without ever checking the *dogs bark* condition. To understand how Python parses more complex logical statements, let's look at another example. The statement

$$(Cats\ meow \wedge dogs\ bark) \otimes (Fish\ swim \wedge \neg birds\ fly)$$

is evaluated in the interpreter as follows. First the parentheses group

$$(Cats\ meow \wedge dogs\ bark)$$

is evaluated. It is `True`, because both statements it contains are `True` (cats do meow and dogs do bark). Since the second set of parentheses can still affect the outcome (a XOR operation always requires us to evaluate both sides),

$$(Fish\ swim \wedge \neg birds\ fly)$$

is evaluated next. The negation on *birds fly* makes this second statement `False` (*birds fly* is `True`, so *NOT birds fly* is `False`). Now we evaluate the XOR operation

$$(A\grave{B})+(\grave{A}B)$$

where A is the result of the first parentheses group, and B is the result of the second parentheses group. We can solve this easily by recursively applying the operator precedence.

$$(1\times\grave{0})+(\grave{1}\times 0)=(1\times 1)+(0\times 0)=(1+0)=1$$

We interpret a `1` to mean the overall statement

$$(Cats\ meow \wedge dogs\ bark) \otimes (Fish\ swim \wedge \neg birds\ fly)$$

is `True`, which matches what we'd expect because XOR requires exactly one of the conditions to be `True` for the outcome to be `True`.

Set Notation

In mathematics, we often want to denote groups of objects rather than individual objects. For example, we might say that all the students who go to a particular school make up a *set*. Each student can occur in the set only once.

We could denote the set of students S. (I'll follow the convention of using a capitalized English variable name for sets, with a few exceptions.) This defines what items are, or are not, considered to be part of the set. This is often called a *membership rule*. Now that we have a definition of membership, we can denote what items i are in the set ($i \in S$) or, of course, not in the set ($i \notin S$). In this example, $i \notin S$ is syntactically equivalent to saying, "All things that are not students at the defined school." When dealing with multiple sets, we may be interested in which set contains an item ($S \ni i$). Assuming we created a set for each school in a district, this would be equivalent to asking, "What school does a given student attend?" Table 2-2 describes the symbols used throughout this text.

Table 2-2: Set Notation Examples

Notation	Meaning	Example
$i \in S$	Boolean: Item *i* in set *S*.	3 in the set of odd integers is `True`. 4 in the set of odd integers is `False`.
$i \notin S$	Boolean: Item *i* not in set *S*.	3 not in the set of odd integers is `False`. 4 not in the set of odd numbers is `True`.
$S \ni i$	Boolean: Set *S* contains item *i*.	Set of odd integers contains 7 is `True`. Set of cars contains a bicycle is `False`.
$A \cap B$	Product: $i \in A \wedge i \in B$. All items in set *A* that are also in set *B*.	Will produce a set of all items that meet both *A* and *B*. If *A* is the set of all animals and *B* is the set of all mammals, a cat could be included in the product because it's in both the set of animals and the set of mammals.
$A \setminus B$	Product: $(i \in A \neg i \in B) + (i \in B \neg i \in A)$. All items in set *A* that are not in set *B* plus all items in set *B* that are not in set *A*.	All topics from security that do not involve medical tech, plus all topics from medical tech that do not involve security.
$A \cup B$	Product: $i \in A \vee i \in B$. All items in set *A* or in set *B* or in both sets.	All places that are gas stations or grocery stores or both a gas station and a grocery store.
$p \subset S, p \subseteq S$	Boolean: Items in subset *p* are all in the superset *S*. If *p* can contain all the members of *S*, we use $\subseteq$.	A subset of door locks is from the superset of all security devices is `True`.
$p \not\subset S$	Boolean: One or more items in subset *p* is not in the superset *S*.	[Pear, Grape, Rock] is not a subset of foods is `True`.
$\forall p \in S F$	For all items *p* in *S*, apply function *F*.	For all the people in the room, say "hello."
$\binom{n}{k}$ or ${}_nC_k$	The number of unordered combinations of *k* elements drawn from a set of *n* elements. Read "from *n* choose *k*."	From [Ron, Tom, Ann] choose two names: $\frac{n!/(n-k)!}{k!} = \frac{3!/(3-2)!}{2!} = 3$

Finally, there are a few reserved sets that come up so frequently that they've been given standardized symbols throughout most of the literature. You can find these listed in Table 2-3.

Table 2-3: Reserved Sets

Notation	Membership rule
$\varnothing$	An empty set. Often seen as a parameter in algorithms (breadth-first search, for example).
$\mathbb{Z}$	Set of all integers (1, 2, 3, . . .) $-\infty$ to ∞. Can also be used for a subset of integers with a defined membership parameter (such as all multiples of 3).
$\mathbb{R}$	Set of all real numbers (0.25, 1.0, 2.3, . . .) Often just 0.0–1.0 with a scaling factor.

NOTE *This is only one way of annotating sets. Another common method is known as* set generator notation (SGN). *Python supports SGN syntax in a few different ways, the most common of which is the* `range` *function, which takes a starting point, a step size, and a maximum value and returns a set (technically a list) of integers between the starting point and the end point spaced out by the step size. The default starting point is 0 and the default step size is 1, so the only required parameter is the maximum value for the range. Calling* `range(10)` *will return the set of integers between 0 and 9. You'll see this syntax used often in* `for` *loops. We'll make extensive use of the set notation symbols defined in Table 2-3 while dealing with graph theory in the next chapters.*

Attribute Characters

Special attribute symbols are another case where mathematicians love to overload the meanings of symbols. They're used in formulas to denote special conditions for variables, function outputs, sets, and the like. They're also used to distinguish related variables that may share a letter (which suggests that they're related). For example, $(y - \hat{y})$ is often used to denote the difference between the actual and predicted values of some regression function. The variable $\hat{y}$ holds the predicted value, and a difference closer to 0 denotes a more accurate prediction. This notation allows us to intuitively understand that the variables are related to each other, while distinguishing the one of particular interest with the attribute character.

When an attribute character is used in this text, it will be accompanied by a description of its intended meaning in that context.

Greek Letters and Functions

Finally, let's discuss the use of the Greek alphabet to denote different variables, functions, and so on. I mentioned one of these symbols already, theta (θ), and how its interpretation is context-driven. You're probably already familiar with some of the others as well. Some will be used fairly consistently, such as pi (π), which will always denote half the circumference of a circle. Others, such as alpha (α), will be used more liberally. To maintain clarity, we'll discuss the meaning of these symbols in the context of each formula that includes them. Table 2-4 outlines some common functions that are also denoted using symbols for shorthand.

Table 2-4: Function Notation Examples

Name/symbol	Common interpretation	Example
ABS $\|A\|$	Absolute value of *A*. Can also be the length of a vector or array (the number of items contained within the array).	$\|[a,b,c]\| = 3$ or $\|3 - 5\| = 2$
SUM $\sum_{i=0}^{j} F$	Sum of performing a function *F* some number of times from *i* to *j*.	$z = \sum_{i=1}^{4} 2 + i^2 = 20$
PROD $\prod_{i=0}^{j} F$	Product of performing a function *F* some number of times from *i* to *j*.	$z = \prod_{i=1}^{5} i + (i-1) = 105$

Summary

Learning to interpret the symbols and remembering how to apply them is the most daunting part of proof-of-concept engineering. From here on, the actual math we'll be doing is quite simple. If you've completed an algebra course and know how to perform addition, subtraction, multiplication, and division, the coming projects will be no problem for you to understand.

In the next chapters, we'll begin discussing the important fields of mathematical theory and building projects to prove their usefulness. Each theory chapter and the accompanying projects are meant to illustrate some of the theories security researchers can benefit from applying in their tool development immediately. They aren't meant to be a comprehensive treatise on any one theory or any particular security topic. I hope that, by the end of the projects, you begin to see the vast number of ways in which programming applied mathematical concepts can impact your work in security and your daily life.

PART II

GRAPH THEORY AND COMPUTATIONAL GEOMETRY

3

SECURING NETWORKS WITH GRAPH THEORY

Graph theory is a powerful but often overlooked tool in a security analyst's arsenal. A *graph* is a mathematical structure that shows relationships (called *edges* or *connections*) between things (called *nodes* or *vertices*), and graph theory provides a suite of algorithms for analyzing the structure and importance of these different, often interlinked, relationships. As technical a topic as security is, at its core it's about relationships: between computers and networks, users and systems, pieces of information, and so on. By modeling a computer network or social network as a graph, you can examine the composition of the relationships to determine, for example, which computers are integral to a business's communications, or which employees are most likely to forward a spam message, and to whom. Knowing which nodes (machines or employees) pose the greatest risk allows you to intelligently distribute your security resources.

This chapter starts by discussing the diverse applications of graph theory to information security, then goes over the theory itself. We'll cover types of graphs, how to create them efficiently in Python, and some interesting measurements you can perform on them. Chapters 4 through 6 then walk you through applying what you've learned here to analyze computer and social networks, the two types of network you'll face most often as a security engineer. We'll answer questions like which computers in a network received the most data, which members in a group are most influential, and how quickly information is likely to spread through a social network.

Graph Theory for Security Applications

Before we discuss how we can apply graph theory in practice, let's take a look at a simple travel graph example in Figure 3-1.

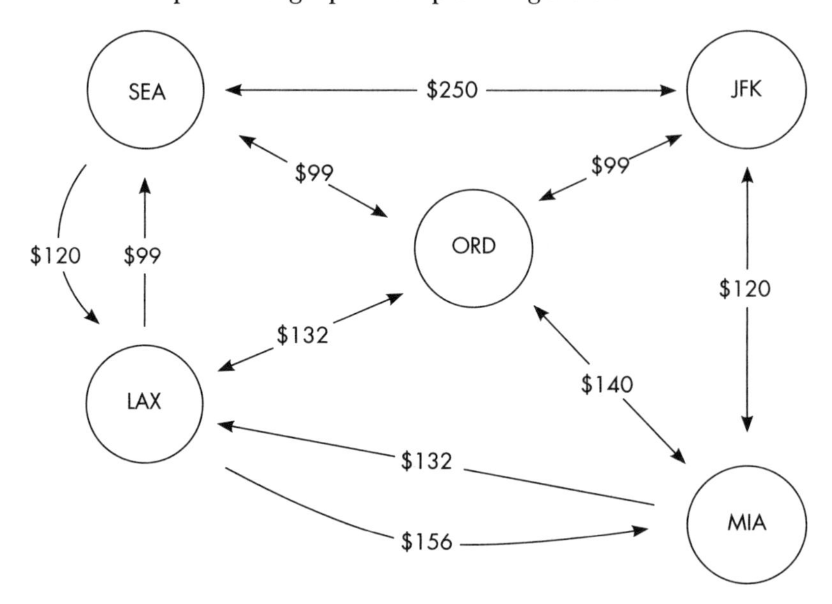

Figure 3-1: A travel graph

As mentioned previously, a graph is defined by nodes and edges. In this example, the nodes (the circles) represent airports in major cities, and the edges (the arrows) represent the cost of a plane ticket between two cities. A graph like this could save you money as you plan a trip. For example, if you wanted to travel from Seattle to New York, you could fly from SEA to LAX, then to MIA, and finally to JFK for a total cost of $396.00. You could also fly from SEA to ORD and then to JFK for $198.00, or directly from SEA to JFK for $250.00.

I don't know about you, but when I fly I don't think just about the cost; I also consider the travel time. In addition to the cost of each potential trip, you can also use this graph to determine the smallest number of stops between any two cities. Fewer stops means shorter trips. As you can see, even a simple graph can contain a lot of information.

When analyzing a computer network, the first step for both attackers and defenders is to get "the lay of the land." That means they can't start to attack or defend anything until they've built a graph of what's available around them. One way to create such a graph is to define computers as nodes and the network connections as edges; this is typical of most network maps. In one of the upcoming projects, we'll model a computer network from a raw packet capture. In this definition, the nodes will be the individual computers, and the edges will indicate when one machine sends packets to another.

Similarly, analyzing a social network can reveal key people and relationships, like the employees who will forward spam messages, or important members of a criminal organization. You can use that information to target or protect members of the network (depending on your job). For example, the FBI uses undercover agents to get information on organized crime families, then builds a social network graph, determines the key figures, and attempts to arrest them. Now, with the prevalence of social media, any amateur sleuth with a laptop can build an alarmingly accurate graph of an organization's (or individual's) social network and use that information to target key members for their own means.

Researchers also apply graph theory, using it to map technologies like cellular networks and cloud computing. For example, scholars have presented ways to apply shortest-path algorithms (similar to limiting the number of airport stops in Figure 3-1) to pick secure routes through graphs representing 5G cellular networks. The research analyzes how messages travel from point to point in the physical layer (OSI model) of the network.[1] We'll use a similar analytical model in Chapter 4 when we graph a computer network from packet data. Other modern research has focused on graphing the logical relationship of components hosted in the cloud. By mapping code usage and typical hypervisor loading activities, scientists have presented a formal way to describe cloud security concerns for virtualization platforms.[2]

Graph theory is also applied to *open source intelligence (OSINT)*, which, in short, collates publicly available information to gain intelligence about a target. An application named Maltego crawls the public web for related terms, email addresses, places, machines, and other details, and creates a graph of where they appear online, like in Figure 3-2. In 2017 at DEF CON, the annual information security convention, Andrew Hay gave an excellent introductory presentation on applied graph theory for OSINT.

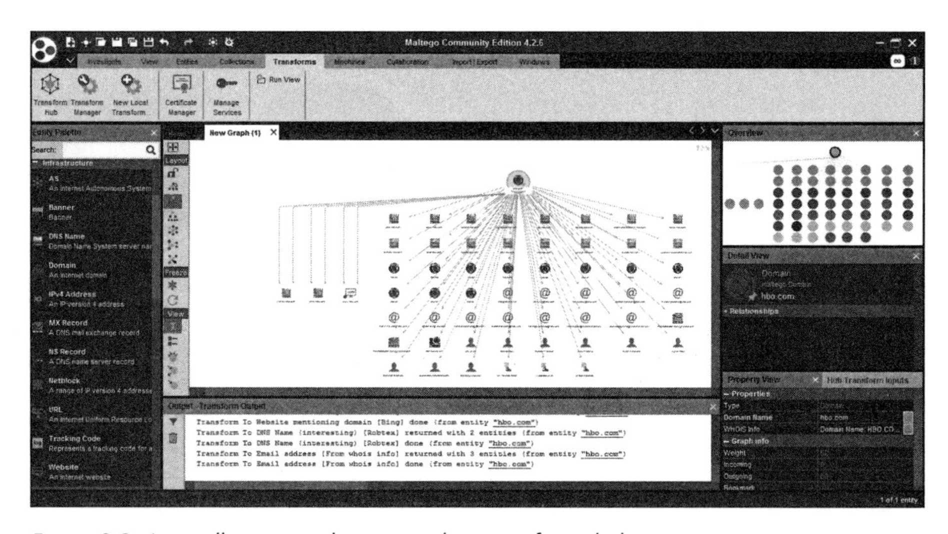

Figure 3-2: An intelligence-gathering application of graph theory

Applications like Maltego blend these logically different types of networks together in one graph, leading to very interesting insights. In one example, my team was able to locate a covert communication channel between two users of different forums. Although the forums were run by different companies, they resided on a shared hosting server. User A joined site X while user B joined site Y. Then, by manipulating the forum software, the two users were able to use local file reads and writes to pass messages on the underlying server. Had my team examined only the social network connections, we would have been stumped, but when we combined information about the social networks and the underlying machine networks, we realized that both accounts could access the same hardware. Of course, you don't need to rely on other people's tools; once you know the inner workings, you can produce your own OSINT-gathering tools, complete with pretty, yet functional, graph displays.

Graphs can also be used to describe how you can go from one condition to another by taking some action. For example, you can go from somewhat secure to completely unsecured by removing the locks from your doors. In this definition, secured and unsecured are called *states*, and removing the locks is the *action* that changes you from one state to another, known as a *transition*. Figure 3-3 shows such a graph, known as a *state machine graph*, that describes the potential for an attacker to move through an environment. Chapter 6 will cover state machines in detail.

You interpret the graph like so: if you're on the internet, and you want to take over an employee's system at your target organization, you can try phishing their customer service team. When you get a willing employee, you send them a remote-controlled malicious payload. You would then be on the employee system, but you might still need to perform some form of privilege escalation to completely take over the system. You can also see that this is just one path you could follow to achieve your objective.

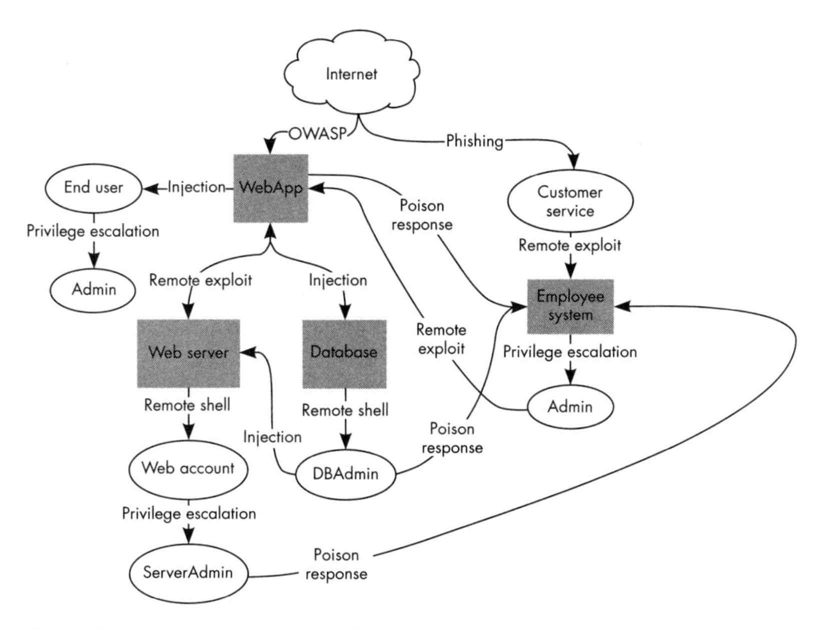

Figure 3-3: A state machine graph

Now that you have an idea of what to do with graph theory, let's discuss the math behind it.

Introduction to Graph Theory

A graph *G* comprises the set of nodes *V* and the set of edges *E*. Information travels between nodes along a set of nonrepeating edges that connect them, called a *path*. A node may forward information to any node it's directly connected to by an edge; the receiving node is the *neighbor* of the sending node. By convention I'll denote an edge as a tuple (u, v) containing an origin node *u* and a terminal node *v*, where both *u* and *v* are in *V* and are unique (not equivalent). We can write this in set notation as follows:

$$E \subseteq (u, v) \in V^2 \wedge u \neq v$$

Sometimes an edge in a graph points back to the same node (breaking my $u \neq v$ assumption); this is known as *self-looping*. For example, if you create a graph of function calls in a program that contains a recursive function, there will be an edge that leaves the recursive function and points directly back to it. Self-loops don't show up too often, but when they do they complicate the analysis and require specialized algorithms, so I recommend leaving them alone until you're very comfortable with the basics of graph theory.

Depending on the type of graph, edges may be bidirectional (*undirected graphs*) or directional (*directed graphs*). If the direction of communication is important to the question at hand, use a directed graph. Otherwise, use an undirected graph. In practical implementation, undirected graphs are usually faster to work with since you assume $(u, v) = (v, u)$. A lot of problem descriptions require directed edges, though. In the travel graph from Figure 3-1, flying from LAX to MIA has a different cost than flying from MIA to LAX, so we need to use directed edges between those two nodes to capture the directional information.

An edge may contain *edge attributes*, additional pieces of information beyond the two nodes it connects. Nodes may also contain additional information called *node attributes*. When one of these attributes is used in ranking a node or edge, it's referred to as the node's or edge's *weight*, and a graph with weights is called a *weighted graph*. In some cases you may even need to add more than one edge connecting two nodes (called *edge multiplicity*) to account for different edge attributes or weights. I'll cover edge multiplicity in its own section later in the chapter, but for now we can extend the travel graph from Figure 3-1 for a simple example. Suppose we find out there are multiple flights leaving SEA for LAX. We may choose to add an edge for each additional flight, along with its cost as the weight. Adding these edges for all city pairs would give us a sense of which airports had the most travel options for our trip. We'll use multiple edges, edge attributes, and weighted graphs to inform our investigations in meaningful ways in the following chapters.

Simple graphs are unweighted, undirected graphs containing no self-looping or edge multiplicity. *Nonsimple* graphs (or, less commonly, *pseudographs*), those that do contain self-loops or multiple edges, comprise the vast majority of interesting graphs you'll encounter in practice.

A *cycle* of a graph G is a nonempty subset of E that forms a path such that the first node of the path corresponds to the last, and no other node is repeated along the path. This is a fancy way of saying a path that forms a closed loop. A self-loop is a special case of a graph cycle with a strict path length of 1. A *cyclic graph* is one that contains at least one graph cycle. A graph that is not cyclic (has no loops) is *acyclic*.

Before we go any further into the theory, let's walk through how to build one of these graph objects programmatically. In the next section we'll go over the current de facto standard library for Python graphs, NetworkX. Having access to the tools in this library will help you construct the examples in this book and play around with the theory at your own pace. The documentation for NetworkX also serves as a great reference to the theory that underlies each function.

Creating Graphs in NetworkX

You can use NetworkX (which contains implementations of most graph algorithms) and Pyplot (a part of the Matplotlib library) to generate and display an undirected graph. Listing 3-1 creates a graph with seven nodes and six weighted edges, then displays it.

```
❶ import networkx as nx
  from matplotlib import pyplot as plt

❷ G = nx.Graph()  # Create the default Graph object
❸ G.add_node('f') # Adds a node manually
  G.add_node('g') # Adds another node manually
❹ G.add_edge('a', 'b', weight=0.6) # Will add missing nodes
  G.add_edge('a', 'c', weight=0.2) # and connecting edges
  G.add_edge('c', 'd', weight=0.1) # Weight is one type of edge attribute
  G.add_edge('c', 'e', weight=0.7)
  G.add_edge('g', 'c', weight=0.8)
  G.add_edge('f', 'a', weight=0.5)
❺ pos = nx.layout.spring_layout(G, seed=42) # Try to optimize layout
  nx.draw(G, pos, with_labels=True, font_color='w')
  plt.show()
```

Listing 3-1: Creating a basic weighted undirected graph

First, we import the two libraries required to build and display the graph ❶. (Aliasing NetworkX as `nx` and Pyplot as `plt` is a common convention in examples online.) Then, we create a basic undirected graph with the NetworkX `Graph` constructor ❷. Defining a graph in this way returns an empty graph (with no nodes or edges).

To manually define the graph's structure, or *topography*, we can add either nodes or edges. To add a node to the graph, we use the `graph.add_node` function ❸ with an argument to use as an identifier (ID) for the node (during lookups, for example). In this case, the ID is the string literal `f`, but an ID can be any object that could act as a key to a Python dictionary (tuples, for instance). The `graph.add_edge` function, which takes two nodes and optional edge attributes as arguments, adds edges directly to the graph ❹. If either `a` or `b` (or both, as in this case) doesn't exist in the graph, NetworkX will helpfully add the missing node(s) before adding the edge. With directed graphs, the order in which you pass the nodes to `graph.add_edge` specifies the edge's direction: the edge starts at the first node and concludes at the second.

The real strength of graphs lies in their visual interpretations, as humans very often can detect patterns in information visually that they wouldn't have found otherwise. NetworkX supports several options for displaying graph information, including Matplotlib and Graphviz. For this example, we lay out the graph with one of NetworkX's built-in layout functions, `nx.layout.spring_layout` ❺, which uses a physics model of spring motion to position the nodes. The nodes' initial positions are randomly generated, but you can pass in the `seed` argument to make the image reproducible, which can be important if you want to share the conclusions from your research with others. The resulting node positions are stored in the dictionary `pos`, with structure `{node ID: (`*`x-coordinate, y-coordinate`*`)}`. The function `nx.draw` creates a plot object using these node positions, and Matplotlib displays the resulting figure. The additional parameters to `nx.draw`, `labels` and `font_color`, control the look of the graph, shown in Figure 3-4.

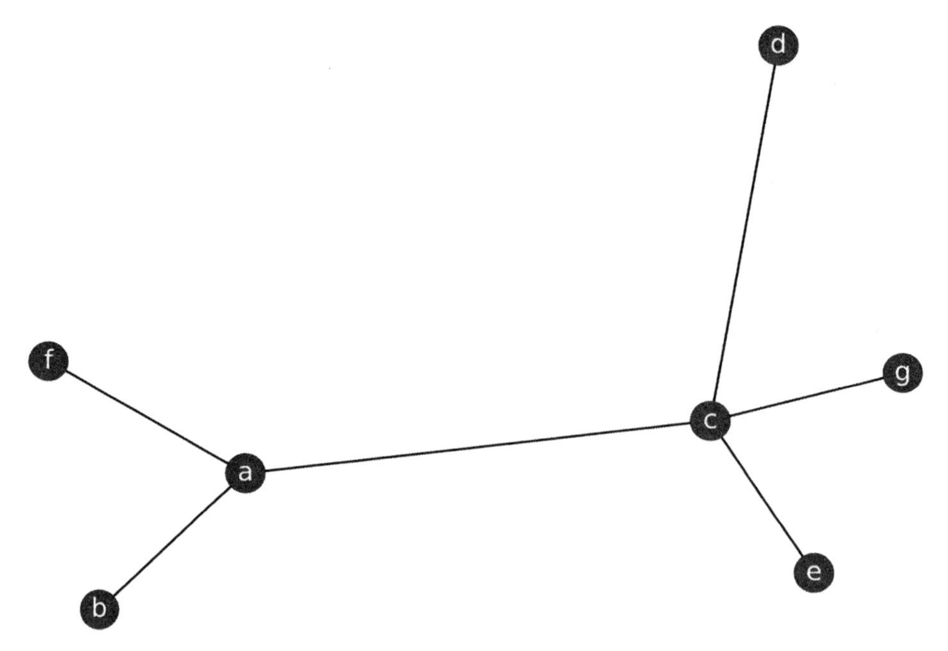

Figure 3-4: An undirected graph

If you remove the `seed` parameter and rerun the code your graph may look different, but it's guaranteed to be mathematically equivalent to the one in Figure 3-4.

Now that we have a way to codify and visualize graphs, let's look at some interesting measurements you can use in your analysis.

Discovering Relationships in Data

In this section, we'll examine a few of the most-used graph properties that can give us insight into the underlying relationships in our data. These properties are expressed as statistical relationships, such as the ratio of the number of possible paths between two nodes to the total number of paths in the graph. Typically we're interested in learning things like which nodes are isolated from other nodes, what are the shortest or longest possible paths between nodes, and how many different nodes can be reached from a particular starting node. There are dozens of possible graph properties to explore, but some are suitable only for certain types of graphs, while others are specialized use cases of these more general properties. The ones described here will give you everything you need to understand the projects in the next three chapters, but it's by no means a complete list.

Measuring Node Importance

A key concept in security is measuring the importance of different assets, be they human or machine, and the impact that compromising them may have on the operation of the organization as a whole. To do so, we need a way to measure which nodes are in critical positions. *Closeness* measures the connectivity of two nodes relative to the other connections in the graph.

NOTE *Closeness can have many interpretations. In Chapter 11 we'll look at how to use physical closeness (the distance between two objects in a physical space) to plan incident response. By selecting security personnel who are already in the vicinity of an incident, you can drastically reduce the reaction time in a lot of cases.*

When you apply closeness across all nodes in the graph, you're measuring some type of *centrality* (roughly, "importance") for each node. There are several types of centrality defined for graphs. The proper one to use depends on the behavior and structure of the network you're trying to analyze.[3] Sometimes you won't know in advance which measure of centrality makes the most sense for your problem. In these cases, start with simpler metrics (like closeness centrality) and move on to testing with other, more complex ones. We'll cover two types of centrality: betweenness centrality and degree centrality.

Finding Nodes That Facilitate Connections

Betweenness centrality considers nodes that connect other nodes together as more central to the graph. Consider a computer network like the one in Figure 3-5, where some systems act as proxies to connect users to databases.

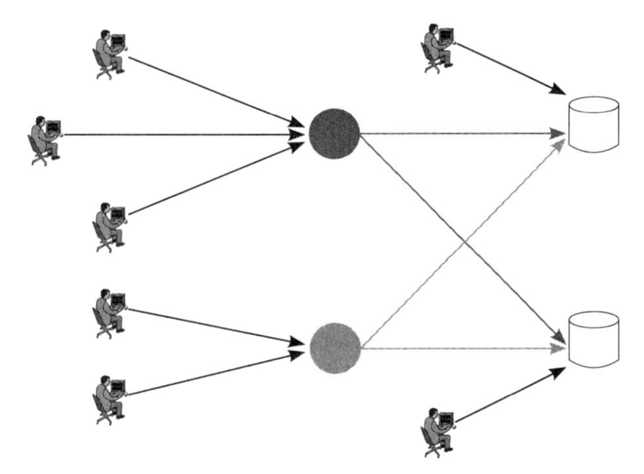

Figure 3-5: A simple proxy network

Betweenness centrality rates the gray proxy nodes much higher than any of the other nodes (the users and databases), since five of the seven users must connect to one of the two proxies to reach either database. The light gray circle at the top is between six paths (3 users × 2 databases = 6 paths) and the dark gray circle at the bottom is between four paths (2 users × 2 databases = 4 paths). The centrality is further strengthened by the fact that these five users must pass through their respective proxy to reach the databases.

Formally speaking, betweenness centrality of a node u is the sum of ratios of all shortest paths from node s to node t, which pass through node u (noted as $\sigma_{(s,t)}(u)$), compared to the total number of shortest paths between node s and node t (noted as $\sigma_{(s,t)}$) for all paths where $s \neq u \neq t$. Putting it all together looks like this:

$$Betweenness\ (u) = \sum_{s \neq u \neq t} \frac{\sigma_{(s,t)}(u)}{\sigma_{(s,t)}}$$

The betweenness scores can be normalized to the number of nodes in G. The normalization function is $2 / ((n - 1)(n - 2))$ for undirected graphs, and $1 / ((n - 1)(n - 2))$ for directed graphs (where n is the number of nodes in the graph). The difference is due to the impact of directionality on the normalization. For undirected graphs, adding an edge between two nodes affects the betweenness score of both nodes and therefore carries twice as much influence as the same edge in a directed graph that impacts only one node (the source node). The normalization scores for both a directed and an undirected graph with five nodes each are calculated as follows:

$$undirected = \frac{2}{((4)(3))} = \frac{2}{12} = 0.166$$

$$directed = \frac{1}{((4)(3))} = \frac{1}{12} = 0.083$$

Unlike some other measures of centrality (such as closeness centrality), normalizing betweenness centrality is optional in NetworkX, and is specified by the Boolean keyword argument `normalized=True`. Listing 3-2 shows how we can retrieve the betweenness centrality scores for the map generated in Listing 3-1.

```
b_scores = nx.betweenness_centrality(G, normalized=True)
nx.set_node_attributes(G, name='between', values=b_scores)
print(G.nodes["c"]["between"])
```

Listing 3-2: Betweenness centrality for the graph created in Listing 3-1

The normalized result for the example graph should be approximately 0.8. There are a total of 15 shortest paths between all node pairs in Figure 3-4 (excluding pairs with c as the start or end node). Of those 15, 12 paths pass through c at some point (12 / 15 = 0.8). The Jupyter notebook for this example shows how you can manually calculate the betweenness score by looping over the node pairs and counting the number of shortest paths that contain the target node.

NOTE *The mathematical definition of betweenness centrality just shown and the algorithm used by NetworkX to compute the betweenness centrality score are both from a paper by mathematician Ulrik Brandes, which has a lot of useful information on the theory of closeness and betweenness.*

Betweenness centrality has many applications within information security and network analysis because it represents the degree to which a node facilitates communication between other nodes. For example, a node in a computer network with high betweenness centrality would have more control over the network traffic, because more packets will eventually pass through it. For this reason, betweenness centrality can also be used to identify good places to perform inspections on network traffic. Another application is understanding critical points of failure in social networks, which we'll discuss more in Chapter 6.

Measuring the Number of Node Connections

Centrality can also be measured by how many neighbors a node has; this is known as *degree centrality*. Intuitively, degree centrality favors nodes that have a larger number of connections to other nodes in the graph. (Betweenness centrality can be seen as a specific measure of degree centrality.)

For undirected graphs, degree centrality is calculated as the fraction of all nodes that are connected directly to a node u. You'll commonly see the neighbors of a node u annotated as $\Gamma_{(u)}$.

$$Degree(u) = \frac{|\Gamma_{(u)}|}{|V| - 1}$$

Remember from the math primer at the start of the book that the absolute value of a set of nodes ($|V|$, for example) is the same as the number of nodes in the set. We subtract 1 from the length of V to account for the fact that node c cannot be a neighbor of itself. Undirected degree centrality is calculated using `nx.degree_centrality`, as shown in Listing 3-3. The bold areas show the few changes required from Listing 3-2.

```
d_scores = nx.degree_centrality(G)
nx.set_node_attributes(G, name='degree', values=d_scores)
print(G.nodes["c"]["degree"])
```

Listing 3-3: Degree centrality with changes from Listing 3-2 in bold

The output of this code should be approximately 0.66, meaning node c is neighbors with two-thirds of the total number of nodes in the graph. In Figure 3-4 you can see that node c has four neighbors and, excluding c, there are a total of six nodes that could be neighbors of c. That gives us $4 / 6 = 2 / 3 = 0.66$, which matches the `nx.degree_centrality` result.

For directed graphs, the degree centrality measure gets split in two pieces. The first deals with edges leading into a node, aptly named *in-degree centrality*. The second measure deals with edges leading out of a node, called *out-degree centrality*. The calculation for each is the same as for degree centrality, except that it considers only the subset of edges matching the specified direction. We denote these sets of edges as

$$(u \rightarrow) = E_{(u,)}$$

for out-degree and

$$(u \leftarrow) = E_{(,u)}$$

for in-degree. Listing 3-4 creates a directed version of the graph from Listing 3-1, then calculates in-degree and out-degree centrality for each node in the graph.

```
❶ G = nx.DiGraph() # Create the default Graph object
G.add_edge('a', 'b', weight=0.6)
G.add_edge('a', 'c', weight=0.2)
G.add_edge('c', 'd', weight=0.1)
G.add_edge('c', 'e', weight=0.7)
G.add_edge('g', 'c', weight=0.8)
G.add_edge('f', 'a', weight=0.5)
❷ i_scores = nx.in_degree_centrality(G)
❸ o_scores = nx.out_degree_centrality(G)
nx.set_node_attributes(G, name='in-degree', values=i_scores)
nx.set_node_attributes(G, name='out-degree', values=o_scores)
print(G.nodes["c"]["in-degree"], G.nodes["c"]["out-degree"])
```

Listing 3-4: Creating a directed graph to measure in-degree and out-degree centrality

To make the graph directed, we replace the generator `nx.Graph` with `nx.DiGraph` ❶. Then, we use `nx.in_degree_centrality` ❷ and `nx.out_degree_centrality` ❸ to get their respective measures. The result of the code should be 0.33 for both values. If you examine the data, you'll see that node *c* has two incoming edges and two outgoing edges of the six total edges we defined. For each measure, then, the math works out to be 2 / 6 = 1 / 3 = 0.33. If you try running Listing 3-4 against an undirected graph, you'll get an error of the type `NetworkXError`, because in-degree and out-degree are specific to the `nx.DiGraph` and `nx.MultiDiGraph` objects.

NOTE *Several algorithms are implemented only for directed or undirected graphs in NetworkX, so if you use a statistical measurement that isn't supported for the type of edges in your graph, you'll need to implement the statistic for the other graph type yourself. We'll see an example of this in Chapter 6.*

The family of degree metrics allows us to specify the direction of information flow while calculating the scores, whereas both the closeness and betweenness measures make assumptions about the directionality. To understand why this matters, consider analyzing network traffic related to a *distributed denial-of-service (DDoS)* attack. A DDoS attack floods a network or specific target machine with more traffic than it is capable of processing, thereby blocking legitimate users' access. As packets travel from one system to the next, they create directed edges on the graph. A sudden increase of in-degree centrality would be seen at the target nodes, which could allow a script to automatically detect and respond to this threat. By including the direction of information flow, you can often provide more meaningful context to your graphs.

Analyzing Cliques to Track Associations

Investigators use *clique analysis* to track the associations of different groups who aren't kind enough to hand out membership lists. By collecting a list of who is talking to whom (and sometimes when), you can find interconnected clusters, or cliques. In theory, a clique β in graph G is any subset of V in which each node is adjacent to each other node in the set. Think of this as a group of friends who have all met each other, or a cluster of computers that are all connected. Some material may refer to these constructs as *complete subgraphs.* Figure 3-6 shows an undirected graph containing different cliques.

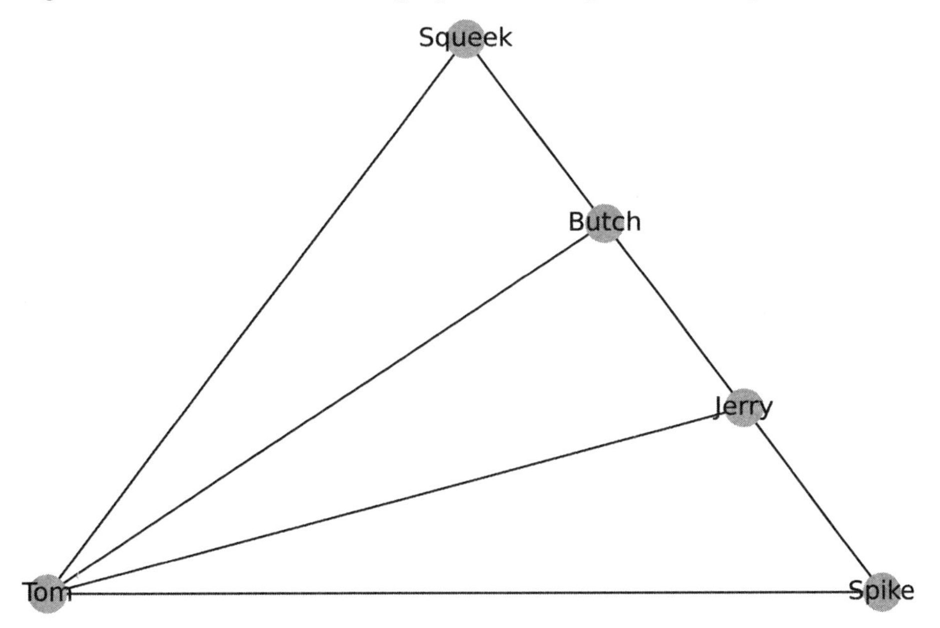

Figure 3-6: A cartoon character graph

A node may be in zero or more cliques. In the graph in Figure 3-6, for example, Tom is in three cliques: Tom, Spike, and Jerry; Tom, Butch, and Jerry; and Tom, Squeek, and Butch. Listing 3-5 creates the graph from Figure 3-6 and calculates a clique membership score for each node.

```
clique_graph = nx.Graph()
clique_graph.add_edges_from(
    [
        ("Tom", "Jerry"),("Butch", "Jerry"),("Spike", "Jerry"),
        ("Spike", "Tom"),("Tom", "Squeek"),("Tom", "Butch"),
        ("Squeek", "Butch")
    ]
)
clq = nx.algorithms.number_of_cliques(clique_graph)
tot = nx.algorithms.graph_number_of_cliques(clique_graph)
for m in clq:
    print(m, (clq[m]/tot))
```

Listing 3-5: Creating the cartoon clique graph in Figure 3-6

The call to `nx.algorithms.number_of_cliques` tallies the number of cliques each node belongs to, which you can use to easily find the node in the most cliques. To find the total number of cliques in the graph, we use `nx.graph_number_of_cliques`. We can then combine the number of cliques for each node and the total number of cliques to create a normalized score to determine which members of a network are key facilitators. The output from running this example code should be:

```
{'Tom': 1.0, 'Jerry': 0.66, 'Butch': 0.66, 'Spike': 0.33, 'Squeek': 0.33}
```

Tom is in every clique, Jerry and Butch are each in two-thirds of the possible cliques, and Spike and Squeek are only in one-third of the possible cliques. Clearly, Tom is the best-known member of this network. In social networks, like a company or an organized crime syndicate, the members in the most cliques are key to facilitating the operations. If we wanted to disrupt the activities of this organization, removing Tom would go a long way toward achieving that. You can also measure clique membership in your network to identify nodes that act as gateways between otherwise separate parts of the network.

The function `nx.algorithms.number_of_cliques` finds the number of *maximal* cliques to which each node belongs—that is, the largest group of nodes that are all connected to one another. In undirected graphs, any two adjacent nodes could be considered a clique, and, in any graph, cliques of four or more nodes contain cliques of three and two nodes, so working with maximal cliques takes those subcliques into account.

You can enumerate all the maximal cliques in a graph with the `nx.find_cliques` function, as shown in Listing 3-6.

```
cliques = list(nx.find_cliques(clique_graph))
print(cliques)
```

Listing 3-6: Creating a list of cliques from a directed graph

The result is a `Generator` object, which is a built-in object type in Python 3. You can either use it directly or cast it to a `list`. We'll see a practical application of finding cliques using `nx.find_cliques` in Chapter 5 when we build a social network graph from posts.

Determining the Connectedness of the Network

Graphs can be connected or disconnected. A *connected* graph is one where every node pair (u, v) has some connecting path (ρ). Therefore, a graph G is *disconnected* if any pair of nodes (u, v) doesn't have a connecting path (ρ) using any subset of edges in E. The only way to know whether or not a graph is connected is to check every pair of nodes to see if they're disconnected. We can write this out neatly using set notation and Boolean algebra:

$$Disconn(G) = \left(\sum_{(u,v)} \rho_{(u,v)} \not\subset E\right) > 0$$

A Boolean statement such as $\rho(u, v) \not\subset E$ returns `1` if it is true and `0` otherwise, so this equation technically counts all pairs of disconnected nodes. In practical implementations, we don't need to continue searching all remaining pairs of G because once we discover a missing edge, we know it's a disconnected graph. We can only say a graph is connected, however, once we've checked every pair of nodes and found no disconnected pairs. You can go through this exercise yourself with the graph from Figure 3-1 to determine if it's a connected or disconnected map.

Figure 3-7 shows the graph from Figure 3-6 extended to be a disconnected graph.

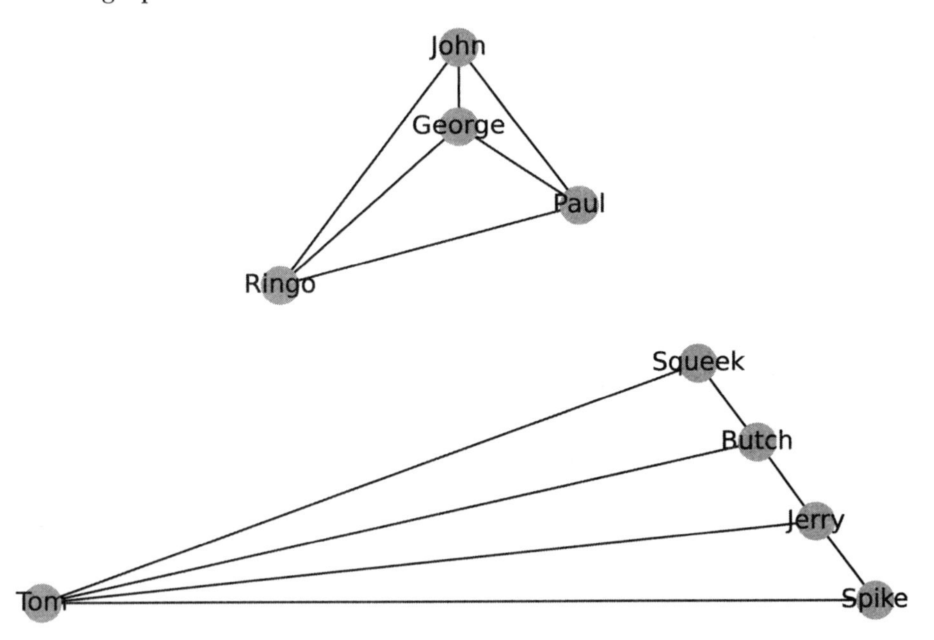

Figure 3-7: A disconnected graph

A disconnected graph is made up of two or more disparate sections called *connected components* (or just *components*). Formally speaking, a connected component of an undirected graph G ($\phi_i(G)$) is a subgraph in which every pair of nodes (u, v) is connected by a path $\rho(u, v) \in E$ (annotated $\rho(\phi_i, u, v)$ for a path in the ith component subgraph). Additionally, none of the nodes in ϕ may be connected to any additional nodes in the superset V. For example, the graph shown in Figure 3-7 has two connected components: one comprising cartoon characters, and the other comprising former bandmates.

NOTE *A clique can never extend beyond a single component, so you'll often care which component a clique is formed in, especially when performing social network analysis.*

Using Graph Edges to Capture Important Details

The final graph property we'll examine is one I mentioned earlier, edge multiplicity. This property is powerful once you know how to leverage the flexibility it brings to your analysis. In many practical instances, like the packet analysis project in the next chapter, there may be multiple edges between nodes that contain valuable information we want to keep for analysis.

For example, graphing the TCP handshake requires multiple directed edges between two nodes. The connecting machine (also called the *client*) sends a synchronization request to the target machine (a SYN packet), which creates one directed edge from u to v in the graph. The target machine then responds with an acknowledgment as well as a request of its own to synchronize (a SYN-ACK packet), which creates a directed edge from v back to u. (In an undirected graph, this response would count as a duplicate of the first edge.) Finally, the connecting machine sends the target machine its own acknowledgment packet (an ACK packet), which creates a second directed edge from u to v in the graph. Figure 3-8 shows two versions of the same graph data containing TCP handshakes between two different groups of systems.

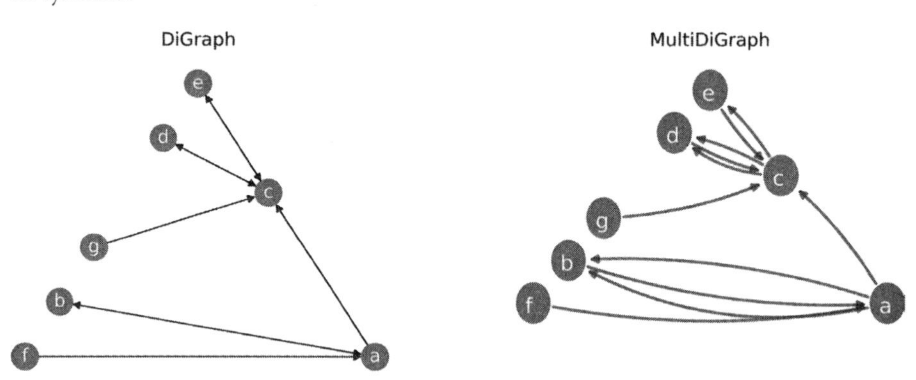

Figure 3-8: Comparing single- and multi-edge graphs

On the left is a standard `DiGraph` representation, which treats repeated communications between two nodes as a single directed edge. Examining this graph alone, you couldn't determine which nodes participated in a TCP handshake. On the right is a `MultiDiGraph` representation of the same data, which retains an edge for each occurrence of communication. Examining this graph, you can easily see that node c initiated a handshake-like exchange with node d. Node a also initiated a handshake with node b.

There are two schools of thought for dealing with edge multiplicity. The first school says you should summarize multiple edges based on their weight ω (and potentially other attributes) into a singular edge, like so:

$$\overline{(u \rightarrow v)} = \sum_{\forall(u \rightarrow v)} \omega$$

If the constituent edges are unweighted, the weight is the number of edges making up the composite:

$$\overline{(u \rightarrow v)_{\omega}} = |(u \rightarrow v)| \in E$$

This summation must take into account the directionality of edges when dealing with directed graphs. (If you're using complex values for edge attributes—such as ranges, which require specialized processing to summarize—you'll be better off implementing your own definition of edge summation in a function within your code.)

The second school of thought is to graph each edge individually and summarize edges only when it comes time to analyze them. Doing so allows you to retain more of the underlying structure. As an example, consider timestamp information on network packets. If you sum the edges into a single edge like the preceding example, you can't see the order in which the edges are created. Retaining each edge allows you to order their creation by timestamp and look for interesting patterns, like call and response pairs in edges.

There's no generally right or wrong way to handle edge multiplicity in a graph. The correct approach is often a little of both schools of thought, as we'll see in the next chapter.

Summary

The power of graph theory lies in the flexible interpretation of nodes and edges. Do nodes represent people, computers, cities, or something else entirely? Do edges measure physical distances or intangible relationships? The answer to all these questions is yes. Be warned, though: this freedom of perception is a double-edged sword. Because there are no strict definitions of what a node or edge represents, you can create a graph whose edges and analysis have no meaningful relationship to reality. An example would be using nodes that represent computers and edges that represent the physical distance between two cities where the computers are located. We typically don't think about how far a message travels on the internet in terms of physical distance, but rather in terms of the number of network "hops" it has to make before reaching its destination. Over the next three chapters, I'll explain the justification for different interpretations of information in more depth, because the meaning of a result, such as closeness centrality, relies on the meaning of an edge's weight and needs a bit of context to make sense.

There's a lot more useful theory than what I've covered here. The book *Introduction to Graph Theory* by Richard Trudeau (Dover, 2001) is an excellent resource.[4] If you're looking for a more advanced discussion, check out *Graph Theory and Complex Networks: An Introduction* by Maarten van Steen (author, 2010).[5] Both books make the topics easy to understand and the math easy to follow. For something more security focused, check out the paper "Applied Graph Theory to Security: A Qualitative Placement of Security Solutions within IoT networks," published in the *Journal of Information Security and Applications* in December 2020, which uses graph theory to analyze the security of IoT network devices and determine suitable locations for monitoring device traffic.[6]

In the next two chapters, we'll put these theoretical concepts to work by examining both a computer network and a human social network to learn which nodes are important to the network, what information is being exchanged, and other important insights about the underlying graph structure. The final graph theory project, in Chapter 6, will give you the tools you need to simulate changes to a network over time. Once you understand the concepts and interpretations, the insights you can gain will make graph theory one of the most powerful and versatile weapons in your analytical arsenal.

4

BUILDING A NETWORK TRAFFIC ANALYSIS TOOL

For our first project, let's start with something familiar. Most of us in the security realm have spent at least some time analyzing packet data and monitoring network traffic. In this chapter, we'll apply the concepts we discussed in the previous chapter—multi-edge directed graphs, centrality, and information exchange—to build our own network traffic analysis tool. We'll use captured network data to build a graph, calculate some metrics to learn about the properties of the observed traffic, and then use centrality measures to figure out what each machine is doing.

When we talk about systems on a network, we often think in terms of their most prevalent use case. Some machines are on a network to serve files, others to route phone traffic, and still others to represent network users. By figuring out what part the machines are playing, we can make an educated guess about the type of traffic to expect from each machine.

We'll use the information exchange ratio to determine which machines are creating and receiving the most traffic of a given type; this will help us determine the usual levels of traffic and thus potential threats. Finally, we'll get started capturing and analyzing network traffic around us with a proof of concept that will generate graphs from live packet capture.

Let's begin by looking at an example network map.

Network Topology Visualization

Most GUI-based packet analysis tools, like WireShark or Zenmap, allow you to visualize the network's topology, combining packet analysis with graph theory to infer information about the network structure. Figure 4-1 shows an example captured on my research network.

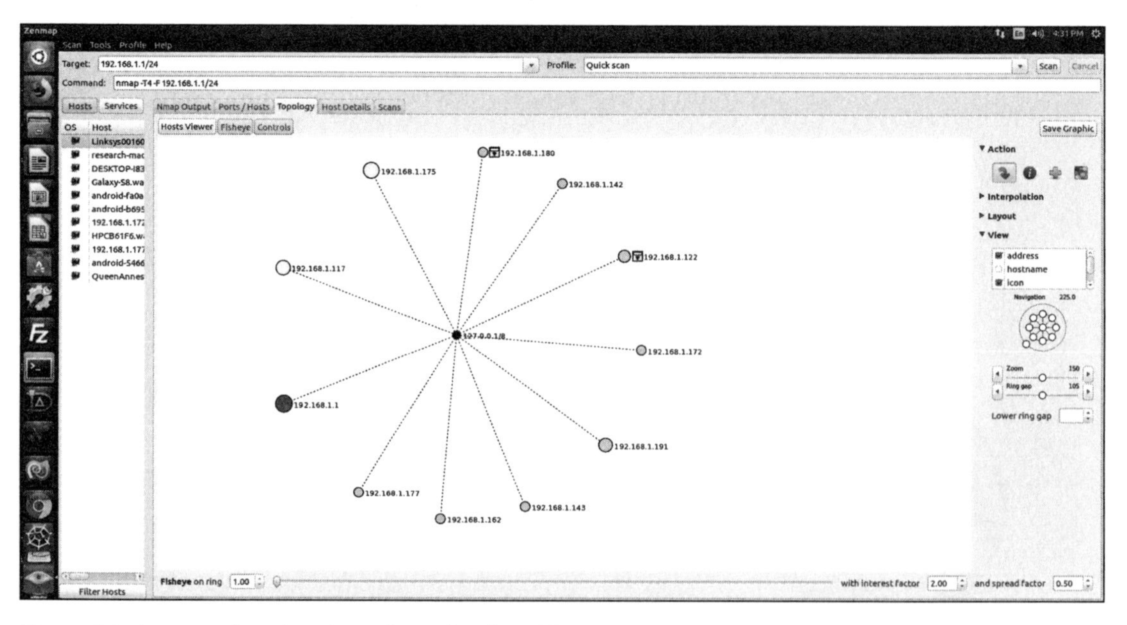

Figure 4-1: An example network topology view from Zenmap

Recall from Chapter 3 that *V* represents all the vertices and *E* represents all the edges; *V* and *E* combine to make the graph *G*. In Figure 4-1, each node in *V* represents a system generating traffic on the network. Each edge in *E* is a communication pathway defined by an observed packet. The nodes and edges both have attributes pulled from the dissected packet fields; we'll use these attributes for further analysis. From the graph of my research network, we can infer that my machine was able to connect with 11 other machines located on the same local area network segment.

Generally speaking, we can interpret this graph as showing the communication relationship between computers on my research network. We can use this relationship map to infer conclusions about expected and unexpected behaviors (like why your coffee pot is sending network traffic to your printer). This can be extremely useful in security systems, as you might expect.

Most traditional network monitoring tools rely on *signature detection* to classify malicious traffic, wherein the monitoring tool will scan for behavior that indicates threats, such as a packet with a sender IP of a known command-and-control server. Typically these signatures take two forms. The first, and most popular, is an *Indicator of Compromise (IoC)*, which represents a unique action taken by malware. As their name implies, IoCs can help identify if a system has been compromised. For example, if a research analyst finds that a new malware variant tried to contact a particular URL during its setup, network administrators can add a rule to their monitoring software that blocks traffic to that URL and sends alerts on a potential infection. The problem is that the IoC approach relies on previous knowledge of behavior that's unique enough that you can identify the infection with a high probability of success and a low probability of false alarms. This behavior can take hours of human research to identify and only minutes for the malware authors to change in their next variant. The sheer number of IoCs to keep up with is staggering, and applying them all—to all network traffic—can sometimes slow things to a crawl.

We can remedy this with the second type of signature detection, aptly named *anomaly detection*. This signature relies on elements of graph theory to create a set of network metrics that are considered "normal" behavior. During live traffic analysis, an operator is alerted if one of these values moves outside the defined range (which will usually include an acceptable variance). By applying graph theory to network traffic, you'll design systems that can detect and react to anomalous traffic without relying on previously seen samples. You can then take this a step further and define a system to automatically respond based on the type of alert being generated.

To get from the theory we've discussed to an anomaly detection system, we have to first figure out how to turn network traffic data into a graph representation we can analyze. We'll need to add another library to the mix to extract the data we want and feed it into NetworkX in a meaningful way.

Converting Network Information into a Graph

We'll use the Python library Scapy to extract information from a packet capture file, known as a *pcap*, and then create a graph from that information using the concepts from Chapter 3. Scapy is Python's version of a Swiss army knife for packet manipulation, providing tools for capturing, analyzing, crafting, and transmitting network packets. Scapy can even be used to quickly define entirely new network protocols. Scapy works off a platform-specific packet capture library. On Linux this is libpcap, which comes installed by default on most modern Linux platforms; it's installed by default on the BSD-based distributions and macOS, and the stable version is usually installed by default on other Linux-based distributions. On Windows you'll need to install an alternative library such as WinPcap (now deprecated) or Npcap (*https://npcap.com*). If you've worked with other packet analysis tools, like WireShark, on your Windows machine, you might already have one of these installed.

We'll read packets from the *network_sim.pcap* file, available for download in the book's GitHub repository. Our aim is to recognize machines on the network that were behaving outside of normal, "expected" behaviors. We're going to analyze the packets to identify the machines present in the data, who communicated with whom, and what type of communication was happening. To do so, we'll apply a bit of knowledge about network protocols and a healthy dose of statistical analysis to our packet graph.

Building a Communication Map

The capture file contains traffic logged by a Snort collection point (*https://www.netresec.com/?page=ISTS*). The capture file contains 139,873 packets from 80 unique *media access control (MAC)* addresses. A MAC address is a unique identifier burned into the hardware memory of your *network interface card (NIC)* by the manufacturer. At a very simplified level, the NIC's job is to physically transmit the data to the next device on the network (usually some type of router or switch). If you're using an Ethernet cable, the NIC will send the electric pulses along the wire. If you're using a wireless NIC, the data will be broadcast via some form of receiver and transmitter combination. When you sign on to a network at home or the coffee shop, your NIC sends its MAC address to the router, which assigns an IP address to the system based on its MAC. If the router has never seen the MAC before, it will allocate the next free IP address, but if the machine has been assigned an IP before and that IP is still available, the router will usually assign the same one again. However, sometimes the previous IP address has already been assigned to a different NIC, so the router will assign a new IP to a MAC it's previously seen.

We'll use the source and destination MAC of each device involved in a packet transfer as the edge identifiers in our graph. It's unlikely that a machine has completely switched its NIC between connections, though, so the MAC address should remain the same. By using the MAC address to identify each machine, we'll be able to recognize the same NIC across different IP addresses and build a somewhat accurate communication map that isn't confused when a machine's assigned IP address changes.

NOTE *This method isn't always accurate.* MAC spoofing *is a technique that overwrites the system's MAC address with a new address. It can be used to impersonate another network device or evade surveillance on untrusted networks. Furthermore, machines often have more than one NIC installed, which means they may connect to the same network with multiple MAC addresses simultaneously. Using MAC addresses to identify systems is fine for our purposes here, but just be aware that in reality it's not guaranteed to be correct.*

Now that we know what data we can use to identify systems, we can focus on the types of network data we're interested in. With nearly 140,000 packets available, we want to filter to reduce the noise in the data and make our processing more efficient. This is where your knowledge of network protocols will come into play. There are potentially dozens, if not hundreds, of different network protocols present in network traffic. By understanding different protocols and when they're likely to be used, you can more quickly zero in on the data of interest. We don't have the space to cover packet analysis in depth, so I recommend you read one of the excellent books listed in the chapter summary to learn how powerful good packet filters can be for security analysis.

The sample file includes packet data with the following protocol makeup:

- TCP: 137,837
- UDP: 2,716
- ICMP: 297
- Other: 1,352

Our analysis will focus on TCP and UDP packets (the two major types of packets used in common network communication like web traffic). TCP and UDP are built above the IP layer, so we'll ignore any packets that don't have an IP layer to filter out all but these protocols. We'll also extract IP addresses and ports. The port numbers will be important as we discuss the type of communication, because a lot of software (like databases and web servers) tend to have default port numbers, so their presence in the packet data can help us guess what systems might be on each side of the communication. By collecting the IP addresses with the information, we can analyze which MAC addresses have been paired with multiple IP addresses. This gives you an idea of how much error could be introduced from just using the IP address as the identifier.

Building the Graph

In Listing 4-1 we load the packet data into a `MultiDiGraph`.

```
import networkx as nx
from scapy.all import rdpcap, IP, TCP, UDP

❶ net_graph = nx.MultiDiGraph()
❷ packets = rdpcap('network_sim.pcap')
❸ for packet in packets:
    ❹ if not packet.haslayer(IP):
          # Not a packet to analyze.
          continue
    ❺ mac_src = packet.src          # Sender MAC
      mac_dst = packet.dst          # Receiver MAC
    ❻ ip_src = packet[IP].src      # Sender IP
      ip_dst = packet[IP].dst      # Receiver IP
    ❼ w = packet[IP].len           # Number of bytes in packet
    ❽ if packet.haslayer(TCP):
          sport=packet[TCP].sport # Sender port
```

```
        dport=packet[TCP].dport # Receiver port
  ❾ elif packet.haslayer(UDP):
        sport=packet[UDP].sport # Sender port
        dport=packet[UDP].dport # Receiver port
    else:
        # Not a packet to analyze.
        continue
    # Define an edge in the graph.
  ❿ net_graph.add_edge(
        *(str(mac_src), str(mac_dst)),
        ip_src=ip_src,
        ip_dst=ip_dst,
        sport=sport,
        dport=dport,
        weight=w
    )
print(len(net_graph.nodes))
```

Listing 4-1: Populating the graph from a pcap file

The net_graph MultiDiGraph variable ❶ will be populated from the pcap file loaded with rdpcap ❷, a Scapy function that reads a pcap file and returns a list of Scapy packet objects in the packets variable. To filter for just TCP and UDP packets, we loop over each packet object ❸ and check if it has an IP layer defined ❹. If it does, we extract the source and destination MAC addresses from the base packet with packet.src and packet.dst, respectively ❺, giving us some edge attributes.

Scapy packet objects store properties for each protocol encapsulated in the packet in layers, with Ethernet card data, like the MAC address, stored in the base layer. We access additional layers with dictionary-like indexing: for example, the source and destination IP addresses from the IP layer are in packet[IP].src and packet[IP].dst ❻. We extract these to store as edge attributes. To weight edges in E using the number of bytes sent per packet, we save the packet[IP].len property ❼ in w, and store that in the edge's weight attribute later. Using weight as the specific attribute name will allow NetworkX to pick it up and use it during analysis. Weighting each edge by the length of the packet's IP layer is a simple way to estimate the amount of data transmitted between machines.

Finally, we check the packet for a TCP ❽ or UDP layer ❾. We need to perform this additional check because not all packets with an IP layer are from the TCP or UDP protocols. For example, Internet Control Message Protocol (ICMP) packets have IP layer information but aren't in the same format as a TCP or UDP packet.

If a TCP or UDP layer is present in the packet, we extract the source and destination ports; otherwise, we skip the packet. We create an edge for each eligible packet using the collected properties as edge attributes and MAC addresses as node IDs ❿. Finally, we can print out the length of the net_graph object, which will tell us that 80 nodes were created. Figure 4-2 shows a 3D graph representation of the network data.

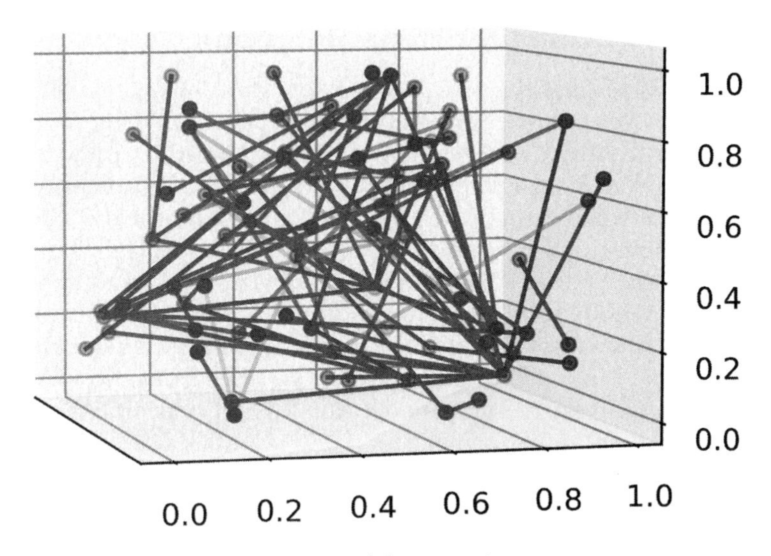

Figure 4-2: A 3D representation of the network

I've generated the axis values for this figure using the `nx.random_layout` function as placeholders for the moment, since we haven't yet defined what we're looking for. The function generates only the *x* and *y* values by default, but you can pass the parameter `dim=3` to have it generate three-dimensional coordinates. We'll be doing the rest of our analysis in 2D, but I wanted to show an example of the graph the way most people think about it—in 3D. Being able to easily display complex networks in 3D like this is a huge time saver. Even though this graph is tangled, you can already get a sense of important nodes in the communication. One node in the upper central area, for example, has many edges connecting it to many other nodes in the graph. Beyond very basic observations, though, you can't gain much insight.

The number of nodes and complex interactions lends itself perfectly to an automated graph analysis approach. Using the theory we've already covered, we'll untangle this graph into organized and informative subgraphs. You'll apply your newfound knowledge of edge filtering and summation to discover which nodes are communicating, using interesting protocols like HTTP and HTTPS. We'll examine which machines are contacting a large number of other machines using a measure of out-degree connections, and finally we'll explore a proof-of-concept program that will allow you to capture and analyze packets from your own network.

NOTE *Extracting only a few properties from each packet keeps the memory footprint of the analysis script low. See the Scapy* `packet` *object documentation for all available properties.*

Identifying Suspicious Machine Behavior

Let's reexamine the concept of closeness in the context of our network data. Given that we recorded the packet's destination port and defined the edge weight as the number of bytes transmitted in the corresponding packet, a natural first question to ask about the network is, "Which machines are using which protocols to communicate?" If we assume that traffic destined for certain ports belongs to a certain protocol or application (like 80 for HTTP, or 22 for SSH), the task is equivalent to asking, "Which node sends the most data (measured in packet bytes) to a given port?" Our simplifying assumption is actually the underlying basis for quick protocol fingerprinting in network tools like Nmap, so I feel comfortable making this particular leap of faith. We can reformulate the question of protocol use more formally as:

> Given a set of protocols Ψ, determine which node has the highest weighted out-degree for protocol $\Psi_{(i)}$.

In fact, investigators examining network operations frequently ask this question when they want to identify machines that are behaving abnormally (outside of the observed average), so it makes sense to automate the process.

Subgraph of Port Data Volume

You can investigate data volume on a given port simply and quickly by first creating a subgraph that contains only edges of type $\Psi_{(i)}$ (for example, SSH) and then measuring the weighted out-degree of each node in the subgraph. Listing 4-2 adds a helper function to the code in Listing 4-1 to create a subgraph for arbitrary port numbers.

```
def protocol_subgraph(G, port):
    o_edges = [(u,v,d) for u,v,d in G.edges(data=True) if d["dport"] == port]
    if len(o_edges) < 1:
        return None
    subgraph = nx.DiGraph()
    subgraph.add_edges_from(o_edges)
    return subgraph
```

Listing 4-2: A protocol subgraph helper function

The function `protocol_subgraph` takes a graph and a port number as arguments, collects all edges representing traffic *to* the port, and creates a simple directed graph. The list comprehension with conditional statement `if d["dport"] == port` prunes the edge set to only the edges of interest. It then creates a `DiGraph` object and adds the pruned edge set with `nx.add_edges_from`. As I mentioned previously, this will also add nodes to the graph. Because NetworkX automatically sums the `weight` attribute of multiple edges between the same two nodes in a `DiGraph`, the `weight` attribute of each edge in the `subgraph` will represent the combined byte count of all packets between two devices.

We can then check the volume of outbound traffic of each node in the returned subgraph using the `nx.out_degree` function. Listing 4-3 shows how to retrieve this information for port 80.

```
dG = protocol_subgraph(net_graph, 80)
out_deg = dG.out_degree(weight='weight')
sorted_deg = sorted(out_deg, key=lambda x: (x[1], x[0]), reverse=True)
print(sorted_deg[0])
```

Listing 4-3: Finding machines with the most outbound traffic for a single protocol

First we call the helper function defined in Listing 4-2 with the `MultiDiGraph` created in Listing 4-1 and the port of interest (`80`, in this example) and then we call the `out_degree` function, which returns the raw count of outbound edges for every node in the subgraph. To change the behavior to return the summed edge weight instead, we explicitly pass `out_degree` the `weight` parameter. Usually NetworkX picks up the `weight` parameter on its own, but for some reason it didn't when I tested my code. Adding an explicit reference to the `weight` attribute solved the problem.

NOTE *The out-degree count is different than out-degree centrality, which takes into account the out-degree of other nodes as well through normalization.*

To find the device that sends the most data on port 80, we sort the result using the `sorted` function. The `key` parameter takes in a function to use when sorting complex objects (like tuples or dictionaries). We pass in a lambda function that takes a tuple of the form (*`node ID, out-degree weight`*) and sorts the items in the order (*`out-degree weight, node ID`*), so the nodes are sorted by out-degree first; if there's a tie, the node's ID is used as the tie breaker. The `reverse` option sorts the items in descending order (the default is ascending). The first item in the sorted list now has the highest out-degree, as you should see in the code's output:

```
('1c:6a:7a:0e:e0:41', 592)
```

Since our goal is to identify interesting or anomalous network activity, such as a sudden increase in network outbound activity on key services like SSH and HTTP, we want to take a list of protocols and determine the node with the highest-weighted out-degree for each. This is equivalent to

$$\forall u \in V : \psi_{(i)}^{u\rightarrow} = \sum_{(i=0)}^{j} (u \rightarrow v)^{\Psi_{(i)}} \omega$$

which defines a $|V| \times j$ matrix (also called a two-dimensional array for you coders), where j is the number of protocols to examine. The entry $\Psi_{(j)}^{(u)}$ holds the summed weight of edges with protocol j for node u.

In Listing 4-4 we again leverage the `protocol_subgraph` function from Listing 4-2 to answer the question "Which machines have the highest weighted communication?" with four popular ports: HTTP, Digital Private

Network Signaling System (DPNSS), the Metasploit RPC daemon default port (also used by Armitage team server), and HTTPS.

```
psi = [80, 2503, 55553, 443]
for proto in psi:
    dG = protocol_subgraph(net_graph, proto)
    out_deg = dG.out_degree(weight='weight')
    sorted_deg = sorted(out_deg, key=lambda x: (x[1], x[0]), reverse=True)
    print(proto, sorted_deg[0])
```

Listing 4-4: Locating machines with the highest outbound traffic for multiple protocols

For each of the port numbers in psi, we create a protocol subgraph dG; then, for each node in the subgraph, we sum the weight attribute for all the out-degree edges. Once all the node weights have been calculated for the current protocol, we sort the scores by weight in ascending order and print out the first item from each result set.

Here's the output of the function:

```
80 ('1c:6a:7a:0e:e0:41', 592)
2503 ('00:26:9e:3d:00:2a', 949)
55553 ('1c:6a:7a:0e:e0:41', 52)
443 ('00:0c:29:ac:42:4b', 678)
```

Each line of output gives us the port number, the node address, and the out-degree score for the node with the most traffic for each protocol. The first thing that should jump out to you, as a security researcher, is that the nodes for port 80 and port 55553 are the same. This is interesting because port 55553 is used by the previously mentioned penetration testing software, and port 80 most often represents unencrypted web traffic. This could indicate a scanner of some kind, probing for unencrypted web content and reporting the data back to a Metasploit server. If I were investigating this network for suspicious users, I would start digging deeper into the behavior of 1c:6a:7a:0e:e0:41.

Another item of interest is that DPNSS traffic on port 2503 may indicate the presence of a *Private Branch Exchange (PBX)*, which is a private telephone network used within an organization. It's possible that 00:26:9e:3d:00:2a is some kind of Voice over IP (VoIP) telephone, but you'd need to investigate further to confirm this hypothesis. VoIP is a fun protocol, because when improperly secured, it allows an attacker to eavesdrop on conversations, inject audio into telephone meetings, reroute or block calls, and otherwise terrorize the attached phone system.

Identifying Unusual Levels of Traffic

To find which nodes receive the most data for a given protocol, we could use the receiving port for the list comprehension in the protocol_subgraph function and measure in-degree instead. The question, then, is how to determine whether the amount of traffic received is normal or suspicious. To do this, we estimate the average amount of inbound traffic on the network by summing the weight of each edge and dividing by the number of edges in the protocol subgraph:

$$\varpi = \frac{\sum_{i=0}^{|E|} \omega}{|E|}$$

If we assume that the traffic for a protocol is normally distributed (meaning most systems would receive similar amounts of traffic for a given protocol), we can compare the detected usage to the average with the *z-score* formula, which scores nodes based on the probability that their difference from the average inbound traffic ϖ is due to normal variance. We can choose how confident we want to be (usually between 80 and 99.9 percent) that the variance isn't by chance. A higher confidence level means more variance will be considered "normal" and fewer pieces of data will be flagged as anomalous, or, put more simply, how extreme the difference in the observed and expected value must be before we're willing to consider it "strange behavior." Listing 4-5 shows how to implement this for the HTTP protocol subgraph.

```
from scipy import stats
import numpy as np
protoG = protocol_subgraph(net_graph, 80)
in_deg = list(protoG.in_degree(weight='weight'))
scores = np.array([v[1] for v in in_deg])
❶ z_thresh = stats.norm.ppf(0.95) # 95% confidence
in_degree_z = stats.zscore(scores)
❷ outlier_idx = list(np.where(in_degree_z > z_thresh)[0])
nodes = [in_deg[i][0] for i in outlier_idx]
print(nodes)
```

Listing 4-5: Identifying outliers using the z-score

We start by importing the stats module from the SciPy library and importing the NumPy library as `np`. Next, we define the protocol subgraph `protoG` by passing the source graph and HTTP port 80 to the `protocol_subgraph` function we defined in Listing 4-2. We then calculate the weighted in-degree using the `protoG.in_degree` function. We use a NumPy array named scores to store the weighted in-degree scores. We next look up the *z* threshold value based on the level of confidence we chose; in this example, we choose 95 percent confidence, which relates to a *z* threshold of 1.645 ❶. This is the number of standard deviations away from the mean we'll use to represent the cutoff between normal data and anomalous data.

With this set, we calculate the z-score for each node in the protocol subgraph using the `stats.zscore` function and save it to `in_degree_z`. The z-scores are centered around 0, so there are negative values representing nodes that have weighted in-degree less than the mean. We're not concerned with systems that have less traffic than average for the moment, so we take only the scores that are greater than the threshold we set using the function `np.where(in_degree_z > z_thresh)`, and we call those scores outliers.

NOTE *If we wanted to know the outliers for both high- and low-traffic machines, we could change the z-score function call to* `np.abs(stats.zscore(scores))`*, which would treat negative z-scores the same as their positive counterparts.*

The result is a nested list of one element, so we take the 0th element, a NumPy array containing the index of values in the `scores` array that are higher than the threshold ❷. We save this to a list named `outlier_idx`. Finally, we convert the indexes to node IDs by looking up each element from the `outlier_idx` in `in_deg`.

We run the code and find two interesting node IDs that we're 95 percent sure have received significantly more traffic on port 80 than the other nodes:

```
['7c:ad:74:c2:a9:a2', '1c:6a:7a:0e:e0:4e']
```

Figure 4-3 shows the protocol subgraph for port 80 using the in-degree measure.

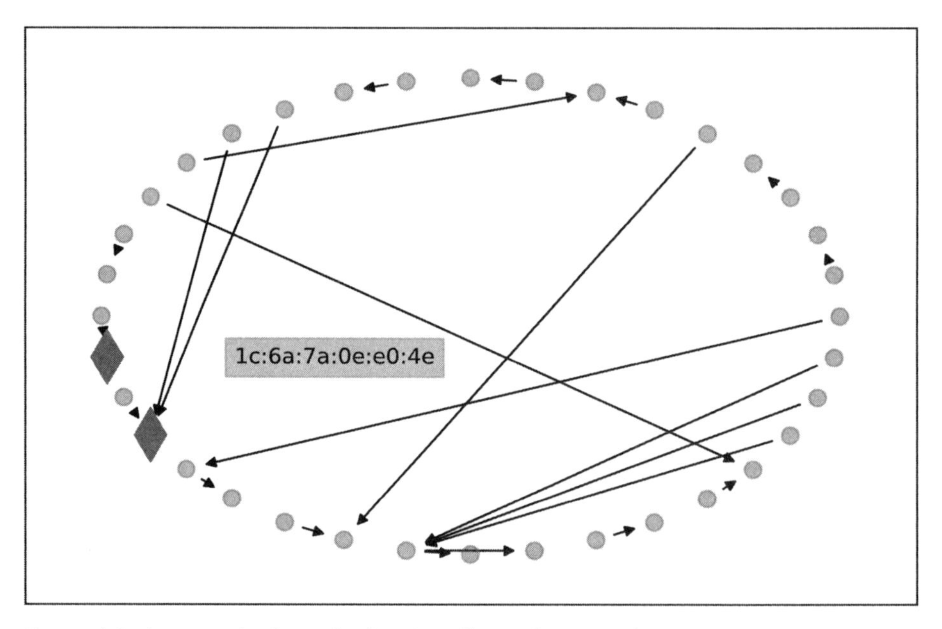

Figure 4-3: A protocol subgraph of HTTP traffic on the network

Each node in this graph sent or received packets on port 80. The black diamond nodes indicate the nodes of interest that were returned by Listing 4-5. The labeled node has the highest in-degree: three distinct inbound connections on port 80. The circular gray nodes were within normal margins for port 80 traffic.

From a security perspective, these two questions (who has sent the most traffic of a given type, and which systems have received a statistically significant amount of traffic of a given type) allow us to assess the behavior of nodes within a network. By measuring the normal, intruder-free traffic over a period of time (say, two weeks) and then comparing it to live captures, you can locate changes in behavior (in traffic volume or actions performed) that may indicate a compromise has occurred. For example, if you know that `00:26:9e:3d:00:2a` is definitely not a VoIP phone, the sudden outbound connections to a telephony network might raise alarms. At the very least, you'd want to contact the machine's operator to understand why this behavior has changed.

Examining How Machines Interact on the Network

As a security analyst, you're likely interested in gaining insight into how machines interact on the network in different but related ways. You may ask protocol-agnostic information, such as "Which machine contacted the most other machines?" or "Which machine absorbs the most information?" On my network, it's normal for there to be very little cross-talk between different machines (for example, my 3D printer shouldn't be talking to my security camera controller) with the exception of my node, which regularly connects to all these machines. By examining the neighbors in my network, you'd quickly see which node I was running. And you might have guessed that I'd also score pretty highly for betweenness centrality because it implicitly favors machines that are neighbors of a high number of nodes that aren't themselves interconnected.

Measuring how much information is exchanged between systems is another way to identify machines trying to exfiltrate data from a network and what systems they're stealing data from. In an information exchange analysis, you're likely to locate machines that serve as information repositories (such as file servers and databases), which typically take in more information than they send out. On the other side of the spectrum are data streaming servers, which produce far more data than they receive. To begin this analysis, first we'll rephrase our questions more formally and then we'll develop the code to investigate them.

Identifying the Solicitor Node

The first question can be stated more formally as:

> For all nodes in *G*, find the node with the highest number of outbound neighbors.

I call this node the solicitor because it behaves like a person going door-to-door in a neighborhood, trying to sell something or collect signatures. Network scanners (like Nmap) will create outbound connections to any machines it can find on the network, making a machine running one of these tools stand out during our analysis. We can find the answer to our question by summarizing any multiple edges between two nodes into single edges, then counting the out-degree of each node, as shown in Listing 4-6.

```
dG = nx.DiGraph()
dG.add_edges_from(net_graph.edges(data=True))
out_deg = dG.out_degree()
out_deg = sorted(out_deg, key=lambda x: (x[1], x[0]), reverse=True)
u, score = out_deg[0]
print(u, score)
```

Listing 4-6: Finding the machine with the most outbound connections

We add all the edges from our `MultiDiGraph` `net_graph` to a new digraph, `dG`, so that NetworkX summarizes any multiple edges between nodes into a single edge with a combined `weight` attribute. Then we use out-degrees from the summarized graph to find the node(s) with the maximum values by sorting the list in descending order and selecting the first node. As I mentioned before, ties will be sorted based on the alphabetical sorting of the node ID. We create a network graph from packets broken by the lexicographical sorting of node IDs.

The code in Listing 4-6 will identify node `1c:6a:7a:0e:e0:44` as the one with the most outbound connections, connected with 13 other nodes in the network. The code in the Jupyter notebook *Chapter 4 - Packet Analysis with Graphs.ipynb* (in the supplemental materials) will collect these machines into a subgraph like the one shown in Figure 4-4.

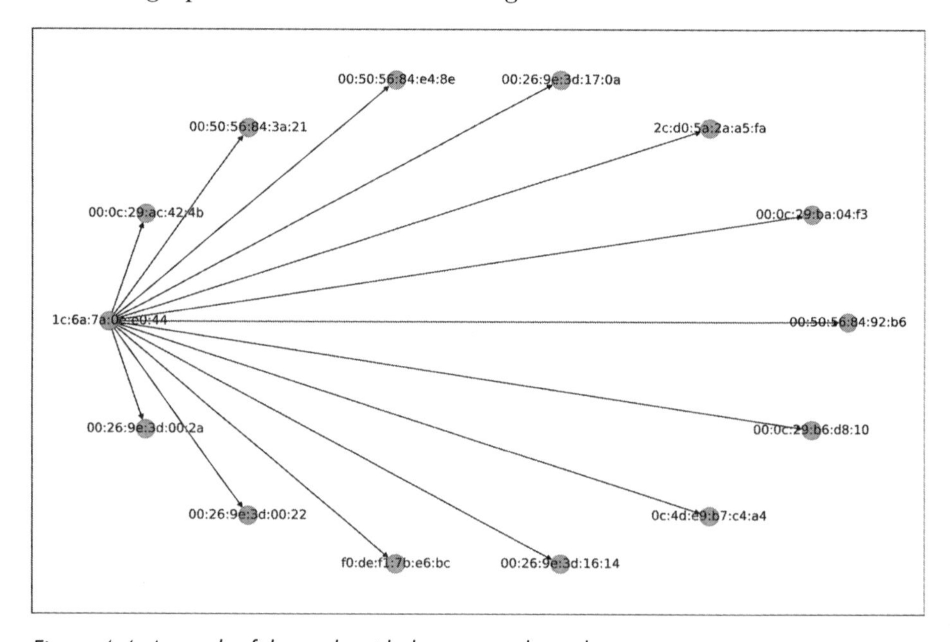

Figure 4-4: A graph of the node with the most outbound connections

You can see the node generating the traffic on the left, with outbound arrows and all 13 systems it has communicated with spread out in what's known as a *shell layout* (because it looks like a seashell).

You'd often perform an analysis like this as a follow-up step after you've identified a suspicious machine. By examining the types of systems that machine has contacted, you can gain more insight into the skills, motivations, and tools of the potential attacker. If you were to continue this analysis, the next step would be to gather the packets associated with each edge and analyze them using your favorite packet analysis methods.

Identifying the Most Absorbent Node

Next we want to find the machine that is absorbing (taking in more than it's sending out) the most data, meaning the node with the largest information exchange ratio. The *information exchange ratio (IER)* can be stated mathematically as the ratio of in-degree weight to out-degree weight for a given node:

$$IER(u) = \frac{1+\sum E_{(u\rightarrow)}\omega}{1+\sum E_{(u\leftarrow)}\omega}$$

Intuitively, a machine that receives three bytes for every one byte sent will get a ratio like 3:1. A machine that generates more packets than it consumes would have an inverse of this ratio, 1:3 (one byte received for every three bytes sent). The formula adds 1 to the numerator and denominator to avoid any 0s in the calculation. NetworkX doesn't provide a handy function for information exchange ratios, so we create the function in Listing 4-7 to calculate the ratio for every node.

```
def exchange_ratios(G):
    res = []
  ❶ for u in G.nodes.keys():
      ❷ out_edges = G.out_edges(u, data=True)
        in_edges = G.in_edges(u, data=True)
        if len(out_edges) > 0:
          ❸ out_w = 1 + sum([d["weight"] for u,v,d in out_edges])
        else:
          ❹ out_w = 1
        if len(in_edges) > 0:
            in_w = 1 + sum([d["weight"] for u,v,d in in_edges])
        else:
            in_w = 1
      ❺ ier = in_w / out_w
        res.append((u, ier))
    return sorted(res, key=lambda x:(x[1], x[0]))
```

Listing 4-7: A function for calculating all IERs

We start by looping over all of the node IDs in the input graph ❶ and calling the `graph.out_edges` and `graph.in_edges` functions for the current node ❷. For nodes with an outbound-edge count greater than zero, we use a list comprehension to gather the weights and then immediately pass this list of weights to the `sum` function, adding 1 to the sum ❸. We assign a base value of `1` to any nodes with out-degree `0` ❹. (A node with no inbound edges and no outbound edges would get a value of 1 / 1 = `1`.) A node with one inbound edge and zero outbound edges would get a score of 2 / 1, and so on. We repeat the process for inbound edges, then divide the two summed weights to produce the IER for the current node ❺. Finally, we return a list of tuples, sorted by the ratio value in ascending order. To sort by descending order instead, you would use the parameter `reverse=True` in the call to `sorted`.

We can call the exchange_ratios function, as shown in Listing 4-8.

```
❶ ier_scores = exchange_ratios(net_graph)
❷ z_thresh = round(stats.norm.ppf(0.99),3)
❸ ier_z = stats.zscore([s[1] for s in ier_scores])
❹ outlier_idx = list(np.where(ier_z > z_thresh)[0])
❺ ier_outliers = [ier_scores[i] for i in outlier_idx]
  print(ier_outliers)
```

Listing 4-8: Finding the nodes with the highest information absorption ratio

This code is very similar to Listing 4-5, where we measured the z-score of the in-degree for each node. We begin by calling the exchange_ratios function on the previously created net_graph object and storing the resulting list of tuples to the ier_scores variable ❶. Next we define the confidence threshold we'll use for our z-score test ❷. A 99 percent confidence level (rounded to three decimal places) will consider data whose value is more than 2.326 standard deviations above the mean to be outliers. To generate the list of z-scores, we pass in a list containing just the score element of each tuple in the ier_scores list ❸. We use the np.where function to find the indexes where the z-score of the IER value is greater than our defined threshold ❹. Finally, we use the returned indexes to look up the corresponding node and IER score in the ier_scores list ❺. When I run the code, I get this output:

```
[('01:00:5e:7f:ff:fa',18570.0),('ff:ff:ff:ff:ff:ff',35405.0),('01:00:5e:00:00:fb',46026.0)]
```

These three nodes all have IER scores we can be 99 percent sure fall outside the range of normal variance for this network. We can safely ignore the ff:ff:ff:ff:ff:ff result, which is known as the network's *broadcast address*. Programs can send a packet to this address when they want to tell the network to broadcast the packet to all machines on the network. We'd expect the broadcast address to have one of the highest IERs since it shouldn't be generating any traffic of its own. We find that the node with the highest absorption ratio is 01:00:5e:00:00:fb, with 46,026.0 bytes absorbed per byte generated. Another item to note about this result is that the lead node's score of 46,026.0 is more than double the next highest outlier, 01:00:5e:7f:ff:fa (ignoring the broadcast address).

Nodes that absorb a lot of data from the network are interesting for several security reasons. For one, a node with a higher IER than average could be downloading a large number of files; a download like this generally starts with one to two packets sent to request the file and a much larger number of packets received that contain the actual file data. The act of downloading a large amount of data from the network isn't necessarily dangerous itself, but it may indicate someone who is crawling the network looking for sensitive information. It could also indicate an attempt to exfiltrate data after a breach has occurred. Therefore, it's worth investigating the cause for the change in the IER.

> **THE EFFECTS OF FILE SIZE ON IER**
>
> You can estimate the number of packets required to send a file by dividing the file size (in bytes) by the network packet size and rounding up the result. For example, suppose a user downloads a file that is 3,584 bytes (3.584KB) in size. For simplicity we'll assume the network uses fixed-length packets of 1,024 bits (1.0Kb). The header of each TCP packet is 12 bytes (96 bits) long. The packet trailer is another 4 bytes (32 bits long), leaving 112 bytes (896 bits) for the payload. Therefore, the network would need to transmit a minimum of four packets (*FileSize* / *PayloadSize* = 3,584 / 895 = 4) to deliver the file. The more this behavior occurs, the more skewed the IER will become.

I hope at this point I've piqued your curiosity about the structures you can find hidden in the network graph. This is a great starting point for identifying suspicious machines but leaves plenty for you to investigate on your own. For example, earlier we identified the suspicious machine `1c:6a:7a:0e:e0:41`. Can you determine from the data how many different IP addresses this machine has? Perhaps building a subgraph to show their communication over time would give you more insight into their behavior.

You might also try applying the clique analysis techniques we discussed in Chapter 3 to see if you can locate any communication clusters. First, ask yourself, "What network scenario would create a clique between computers?" and then see if any cliques in the network support or disprove this hypothesis. This pattern of guess-then-test is at the heart of all applied sciences. I haven't investigated this avenue in the data myself, so you might find some interesting insights uniquely your own. Gaining these new self-motivated insights is the real power and reward of applying mathematics to security topics.

The Proof of Concept: Network Traffic Analysis

To continue applying graph theory to network traffic analysis, you can download publicly available pcaps or capture packets from the networks around you (with permission, of course!), then build a graph from the capture and analyze to your heart's content. Where many researchers struggle isn't with applying the analysis but with constructing the graph in the first place. You've already seen an example of this in Listing 4-1, so you're ahead of the game! I'll therefore conclude this chapter with a proof that shows how to practically bridge the gap between data in the wild and a pretty graph you can analyze. This proof of concept generates a graph from either a previously captured file or live data read from a network interface, then saves the graph as an edge list file that you can load into other analysis scripts. I encourage you to download the proof of concept, experiment with it, and then incorporate it into a larger tool specific to your application.

You'll need the files *packet_analysis/packet_analysis.py*, which contains the code to define the command line interface (CLI), and *packet_analysis/graph_funcs.py*, which contains the functions we've discussed so far and a few other helpful ones. The `pcap_graph` function defines a function wrapper for the code in Listing 4-1, which allows you to pass in a list of packets to work from. The `save_packet` function is a convenience function used to append captured packet data to a given pcap file using Scapy's `wrpcap` (short for *write pcap*) function. Once the file is written, you can use the `file_to_graph` function to load the captured data using Scapy's `rdpcap` (short for *read pcap*) function. Then you'll use the `pcap_graph` function to convert the packet data into a `MultiDiGraph` object for analysis, as shown in the following code:

```
def file_to_graph(pcap_file):
    packets = rdpcap(pcap_file)
    new_graph = pcap_graph(packets)
    return new_graph
```

Once the graph object has been created, you can use the `save_graph` function (which is a wrapper for NetworkX's `write_weighted_edgelist` function) to write the weighted edge representation out to a file. Storing a packet capture as an edge list reduces the load time for graphs. Rather than converting packets to edges on each analysis run, you simply create the base graph one time and then load it (rather than the pcap data) for future analysis. This workflow is known as *write once, read many* (or *WORM*, after the data storage term of the same name).

Whenever I'm working on a proof of concept, I forgo fancy UIs and usually wrap a CLI around the code I want to test. Keeping things simple lets you focus on the core concepts without getting sidetracked by display issues or unrelated interaction problems. The proof of concept for this chapter uses the optparse library to create a set of packet capture options you can use to configure how many packets to capture, where you want to capture them, and more. To start, open your command console, navigate to the *packet_analysis/* directory, and run

```
$ python packet_analysis.py -h
```

This should bring up the options available for running the proof of concept, as shown in Listing 4-9.

```
Usage: packet_analysis.py [options]
Options:
  -h, --help show this help message and exit
  -i IFACE, --iface=IFACE The network interface to bind to      (required, -i all for all)
  -c COUNT, --count=COUNT Number of packets to capture                        (default 10)
  -r RAW_FILE, --raw-out=RAW_FILE File to save the captured packets to      (default None)
  -s GRAPH_FILE, --graph-out=GRAPH_FILE File to save the created graph to         (required)
  -l LOAD_FILE, --load=LOAD_FILE Pcap file to load packets from
```

Listing 4-9: Proof-of-concept run options

As you can see, the -h option corresponds to help. By default optparse includes this option and will print out whatever help messages you define for each option. These are a great way to jog your memory if you set a project down and have to come back to it. The rest of the options and the logic to implement them are stored in the *packet_analysis.py* file.

To capture some number of packets from all network interfaces, then save them as a graph representation, use a command like this:

```
$ python packet_analysis.py -i all -c 100 -s my_test.edges
```

The -i (--iface) option takes an interface name as a string. In the special case of the string all, Scapy tries to bind to all available network interfaces. The -c (--count) option defines the number of packets to capture before exiting the sniffer. This isn't strictly necessary in your programs, but it helps during prototyping to keep the file sizes manageable. Finally, the -s (--graph-out) option specifies where you want to output the weighted edge list file generated when you call the nx.write_weighted_edgelist function. Once you've saved the packet capture graph as a weighted edge list, you can use it in your own analysis scripts by reloading the edge list into a graph using nx.read_weighted_edgelist:

```
G = nx.read_weighted_edgelist("my_test.edges", create_using=nx.DiGraph)
```

By default, NetworkX creates an undirected graph from the edge list. To create a directed graph instead, you'd pass nx.DiGraph in the create_using parameter.

You can also use the proof of concept to create a weighted edge list file from an existing pcap file, which can come in handy for retrospective analysis. To convert the *net_sim.pcap* file into a weighted edge list file, you combine the -l (--load) argument with the -s parameter like so:

```
$ python packet_analysis.py -l network_sim.pcap -s sim_test.edges
```

The proof of concept also supports capturing packets from a specific interface and saving them to both a pcap and an edge list file. By doing both simultaneously, you retain the most information. You can start future graphs right from the edge list without needing to convert a pcap file first, but can still send the pcap data to other tools. The next command shows how to use the -r (--raw-out) parameter in combination with other parameters to create a pcap file along with the graph. This is most useful when you're capturing from a live traffic stream (otherwise you'd already have the pcap file). To capture traffic, the script needs permission to put the network card into promiscuous mode, which is a restricted function on most systems, so you'll need to run the following command as the root user on Linux and macOS or as the administrator account on Windows. (If you're running your setup in Anaconda, you'll need to create the virtual environment using the privileged account so you can run the script with the proper permissions.)

```
$ python packet_analysis.py -i eth0 -c 100 -s my_test.edges -r cap_test.pcap
```

Make sure you change **eth0** to an interface on your machine when you run this command. If you're on a Windows machine, it can be hard to locate the proper device name. It may be easiest to use the `-i all` option if you're using only one network interface. The result will be an additional file that contains all of the raw packet information.

This method takes up the most storage space of all the options. The exact amount of storage required depends on the number of packets captured as well as the amount of information stored for each edge. Be sure to monitor your machine's storage capacity when doing large captures (more than 2,000 packets, for example). You can set the number of packets captured using the `-c` flag. Alternatively, you can send the resulting files to a cloud storage location, and potentially aggregate captures for truly big data analytics. We'll discuss cloud deployments more in Chapter 13.

Summary

This chapter serves as a good starting point for you to build your future network analysis tools. You should now feel comfortable loading a network as a graph, locating interesting nodes using some statistical analysis, and reorganizing the data to suit your questions. You've seen practical examples of how to capture the source data in the proof-of-concept code. Now it's time for you to expand on your own research. You can add more information to the edge attributes (including time of creation, for example), extend it to handle other protocol layers (ICMP would be a good place to start), and make many other useful improvements. Once you're familiar with turning packet data into measurable graphs and manipulating them using NetworkX, you can refer to research dealing with the structure of computer networks, which is plentiful and easily accessible through search engines, to extend your analysis.

For example, if you're interested in applying graph theory to understand resource usage in the cloud, check out the research paper by Kanniga Devi Rangaswamy and Murugaboopathi Gurusamy, "Application of Graph Theory Concepts in Computer Networks and its Suitability for the Resource Provisioning Issues in Cloud Computing—A Review,"[1] which also contains a list of resources related to graph theory and a description of the theory covered. That section alone makes the paper a must-read in my opinion.

As you've seen through this first project chapter, the key to translating theory into practice lies in formulating well-defined questions. For example, the protocol usage question allowed us to identify a potential threat on the network. Books like *Practical Packet Analysis*[2] by Chris Sanders and *Attacking Network Protocols*[3] by James Forshaw can give you more specific network dissection knowledge that will help you ask better questions of your data. As you read through these books, think of how the tools and techniques you learn about rely on the principles we've applied here. Perhaps you'll find a unique way to analyze a network protocol of interest to you.

In the next chapter, we'll leave behind this world of digital order for the less defined world of social networks and ask questions that will completely redefine our understanding of a graph.

5

IDENTIFYING THREATS WITH SOCIAL NETWORK ANALYSIS

Social network analysis (SNA) is a subset of graph theory that describes complex human interactions mathematically; it can be used in any research where human interaction is a factor. Security researchers use SNA for everything from predicting the spread of malicious content to identifying potential insider threats. We'll do our own SNA in this chapter: we'll build a humble social network graph from Mastodon posts, then use it to understand influence and information exchange among users. Specifically, we'll be looking at a real social network effect known as the *small-world phenomenon*. Then we'll look at how to build a graph from posts, answer a few research questions, and end with a proof-of-concept project, where you'll be able to capture data from your own Mastodon timelines.

The Small-World Phenomenon

Stanley Milgram's *small-world experiments* demonstrated the influence of group conformity on human decision-making. The aim of the experiments was to examine the average path length for social networks in the United States, where the *path length* is the number of people it takes to get a letter from one person in the network to another, seemingly disconnected person. Milgram typically chose individuals in the US cities of Omaha, Nebraska, and Wichita, Kansas, to be the starting points; someone in Boston, Massachusetts, was the typical end point. Upon receiving the invitation to participate, the person designated as the original letter sender (the starting point) was asked whether they personally knew the randomly selected final recipient (the end point). If so, the starting point was to forward the letter directly to the end point. In the more likely scenario—the starting point doesn't know the end point—the starting point was asked to think of a friend or relative who was more likely to know the end point. People along the path could forward the letter to anyone they knew who might be able to get the letter closer to the end point.

Technically speaking, a social network exhibits the small-world phenomenon if any two individuals in the network are likely to be connected through a small number of intermediate acquaintances. Milgram's research showed that our society is a strongly connected network: any two members of the network are likely to be connected through three to six intermediate acquaintances (this is popularly known as *six degrees of separation*, a specific case of the small-world phenomenon). The mechanism at play in the small-world phenomenon is called *preferential attachment*, where a person is more likely to form a connection to someone who already has a lot of connections. Put simply, you are statistically more likely to meet a new person who goes out and meets a lot of new people than to meet a shut-in with only a few social interactions. These types of networks, it turns out, are abundant in nature, having been observed everywhere from animal social structures to the human brain. Clearly, it's worth our time as security analysts to understand it.

Our goal for analyzing our data set of fictional posts and users is to answer the following research questions:

- How much information gets propagated?
- What cliques exist in this network?
- Who are the three most influential users?
- Who are the three most influenced users?
- Who could introduce the most new connections?

Over the rest of this chapter, we'll cover each topic in turn, see how we can reinterpret the previous theory to gain insight into social network user interactions, and explore new graph theory topics, like residual information and node ancestry.

Graphing Social Network Data

To turn our social network data into a graph, first we need to structure it into a searchable table format. We'll be using the pandas library, which gives us access to functions and data structures that help us organize our data to prepare it for graphing.

First, let's look at the raw JSON data. The file *fake_posts.json*, included with the book's supplemental materials, contains 28,034 post-like objects formatted in the JSON schema shown in Listing 5-1.

```
{
❶ "id": 4912964953915055,
  "created_at": "2019-5-22 23:03:22",
❷ "content": "Process within summer especially song when letter nearly.",
  "source": "Data Faker",
❸ "in_reply_to_screen_name": "Some User",
  "in_reply_to_id": 1234334523168,
  "in_reply_to_account_id": 346835683,
❹ "account": {
      "id": "6336091949992",
      "screen_name": "juliekennedy",
      "location": "846 Adam Spring #616\nE Chicago, IL 21342",
      "description": "Faked profile Data",
      "url": "http://www.smith.com/"
  },
❺ "reblog_count": 0,
  "liked_count": 0
}
```

Listing 5-1: A mock API response using an example of the Mastodon schema

The `id` field ❶ holds a numeric ID that the API assigns to an individual record—the post—when the post is created. The `content` field ❷ contains the data being added to the network—that is, the text that makes up the post. Some posts are originals and the rest are replies to posts, or replies to replies, and so on. The `reblog_count` field ❺ tallies how many times a post object received a reply. An object representing a reblog will contain the field names that start with `in_reply_to_` ❸. The `account` field ❹ is a nested JSON object that identifies the post creator.

NOTE *The types and amounts of data available from the Mastodon API depend on the individual user settings as well as the access controls put in place by the instance administrators, so this schema represents the basic template instead of replicating Mastodon's structure exactly.*

Structuring the Data

We'll be retaining a lot more data from the post objects than we did from packets in Chapter 4, and the post object structure is nested, so first we need to load the data file into a pandas `DataFrame` object. A `DataFrame` object is the pandas version of row and column data storage for tabular data. It's similar in structure to a database (and even supports some of the same operations, like filtering and joining data). Using DataFrame gives us a more convenient syntax for sorting and selecting relevant post objects, and it highlights the power of combining analytical libraries. By combining tools (in this case, pandas and NetworkX), you can choose the right tool for a particular job instead of trying to make a library do something it wasn't designed for.

The code in Listing 5-2 defines a DataFrame object from the example data.

```
❶ import pandas as pd
  import json

❷ def user_to_series(dict_obj):
      """Convert a nested JSON user into a flat series"""
      renamed = {}
      for k in dict_obj.keys():
          nk = "user_%s" % k
          v = dict_obj[k]
          renamed[nk] = v
      ret = pd.Series(renamed)
      return ret

  series_data = [] # 1 JSON object per post object
❸ with open("fake_posts.json") as data:
      text = data.read().strip()
      rows = text.split("\n") # JSON objects stored as list of strings
  for row in rows:
    ❹ obj = json.loads(row)   # Converted row string to JSON object
      series_data.append(obj) # Add to JSON list

❺ t_df = pd.DataFrame(series_data) # 1 row per JSON object
❻ post_df = pd.concat([t_df, t_df["account"].apply(user_to_series)], axis=1)
  # Data is flat now. Remove the original JSON object feature.
❼ post_df.drop("account", axis=1, inplace=True)
```

Listing 5-2: Creating a pandas DataFrame object from example JSON data

After importing the required libraries ❶, we define a helper function called user_to_series ❷, which I'll discuss in depth in a moment; at a high level, this function converts each JSON user object into a row suitable for use in a pandas DataFrame. We load *fake_posts.json* in the typical fashion using with open ❸, remove any trailing whitespace characters with the strip function, and split the file data into rows using the remaining "\n" characters. The pandas library can create a DataFrame from a list of JSON objects, so we convert each string row into a JSON object using json.loads ❹ and collect the objects in the series_data list ❺.

Unfortunately, when the JSON contains nested objects, like the account field, pandas doesn't know how to unpack them. We need to turn the nested fields into a flat pandas object using the pandas functions apply and concat ❻ to apply the user_to_series function to each row in the data, creating a flat pandas `Series`. You can think of a *series* as similar to a row in database parlance—it groups all of the data relevant to a single entry.

The `pd.concat` pandas function appends these new features to the current `DataFrame` for all rows. The `axis=1` parameter tells pandas to use the series as *features* (columns in database parlance), which results in the `DataFrame` having a column matching each piece of data in the user field (such as username and ID). Each row then represents the user, and each column holds the value of that field for that particular user.

NOTE *If we didn't tell pandas to use the series as the graph's features, the result would be a pivoted version of the data, where each feature would end up as a row and all the rows would be spread horizontally as features. There are times when you may want this behavior (so store that fact away), but this isn't one of them.*

Lastly we remove the original `account` feature, which is no longer needed, using `DataFrame.drop` ❼. Once we've loaded the initial data set and applied all the column processing, we can print out the structure of the data by calling `post_df.info`. Listing 5-3 shows the structure resulting from Listing 5-2, which you can see in the Jupyter notebook *Mastodon_network.ipynb* in the supplemental materials.

```
<class 'pandas.core.frame.DataFrame'>
RangeIndex: 28034 entries, 0 to 28033
Data columns (total 14 columns):
 #   Column                  Non-Null Count  Dtype
---  ------                  --------------  -----
 0   id                      28034 non-null  int64
 1   created_at              28034 non-null  object
 2   source                  28034 non-null  object
 3   content                 28034 non-null  object
 4   in_reply_to_account_id  10302 non-null  object
 5   in_reply_to_id          10302 non-null  float64
 6   reblogs_count           28034 non-null  int64
 7   favourites_count        28034 non-null  int64
 8   user_id                 28034 non-null  object
 9   user_screen_name        28034 non-null  object
 10  user_location           28034 non-null  object
 11  user_description        28034 non-null  object
 12  user_url                28034 non-null  object
 13  in_reply_to_screen_name 10302 non-null  object
dtypes: float64(1), int64(3), object(10)
memory usage: 3.0+ MB
```

Listing 5-3: Post data structure in pandas

The data structure tells us a few important things. First, the `RangeIndex` property tells us how many rows of data are currently in the `DataFrame` object. In this case we have loaded 28,034 post records, indexed from `0` to `28033`. Next, we can see that there's no longer an `account` column in the list, which means our drop operation in Listing 5-2 successfully modified the `DataFrame`. The number to the right of the column name represents how many rows in the data have a non-null value in that column. We can see most of our columns have values in every row because the non-null count matches the index count. In contrast, the columns starting with `in_reply_to_*` have non-null values in 10,302 of the 28,034 rows. This is because these values are present only on posts that are responses. We'll take advantage of this difference between original posts and replies later.

To the right of the value count is the type of data stored in the column. If you don't explicitly define types for the data as you import it, pandas will do its best to logically interpret the types. Unfortunately, it's really only good at finding integer and float types. For the rest of the columns, you can see it has assigned the generic type `object`. This is the pandas way of saying it really doesn't know what to make of the data in the column. It may be of an unorderable type (like the strings stored in the `user_name` column) or there may be two or more data types in the same column (such as the column `in_reply_to_screen_name`, where some rows have an integer value and others have a null value). Before you begin any analysis, it's important to understand the structure of the underlying data. You'll become familiar with the different data types available and when to use each, but for now we don't need to change anything, so we'll move on to the last two rows of the output. The `dtypes` property just gives a summary of the data types in the column for convenience. We can see that one column was determined to be a floating-point number, three were determined to be integers, and the rest pandas left as generic `object` types.

Finally, the memory usage line estimates the amount of memory used to store the entire `DataFrame` object. You can use this value to get a rough idea of data storage requirements for your application, but there are some caveats here. Depending on your configuration, pandas can calculate this number in one of two ways. By default, pandas simply multiplies the bytes required to store a value of each column's data type by the number of rows in the `DataFrame`. For example, an `int64` value takes up 8 bytes, so the `id` column takes up approximately 8 × 28,034 = 224,272 bytes (a little over 224KB). By repeating this for each column and summing the results, pandas quickly approximates memory usage. The problem is that some data types (the `object` type, for instance) don't have a maximum size, so pandas can only guess the minimum space assigned to these types. That's why there's a + symbol after the memory usage.

Visualizing the Social Network

With the `post_df` object defined, you can analyze the data structure and choose which fields to use in your graph definitions. Let's define a node ID *u* as a unique user account using the network. The `user_id` and `user_screen_name` have a 1:1 relationship, so either is a good candidate for the node ID. The `user_screen_name` field makes the graphs more aesthetically pleasing, but the `user_id` might be better for an automated system—for example, one that uses the network analysis results to look up user profile information by ID. We'll be using the `user_screen_name` field so the graphs are more engaging and memorable; it beats staring at a bunch of randomly generated IDs.

For edges, we'll look at when two users interacted over a post, shown in Listing 5-4.

```
❶ G = nx.DiGraph()
❷ for idx in post_df.index:
  ❸ row = post_df.loc[idx]
  ❹ G.add_edge(
      row["in_reply_to_screen_name"], row["user_screen_name"],
    ❺ capacity=len(row["content"])
    )
print(len(G.nodes))
```

Listing 5-4: Representing the post data as a directed graph

The in_reply_to_* fields allow us to see when a post is in response to an earlier post (ostensibly from another user). When user B replies to a post from user A, we'll consider this an edge between them, $e_{(a \rightarrow b)}$. I'll discuss more about edges and interpreting them as we go along.

First we create a directed graph ❶ from the previously defined DataFrame object.

We loop over each index in the post_df object ❷ and use the DataFrame.loc function to retrieve each row individually ❸. We add edges to the graph whenever one user reblogs another user's message ❹. The user who created the original post (the source node) is held in the in_reply_to_account_id field, and the user who's responding (the terminal node) is held in the user_screen_name field. We then include the length of the text as a specially named version of edge weight called capacity ❺. This is a very simplistic measure of information contained in a post, as we'll discuss more shortly. Finally, we can print out the length of the list of graph nodes to verify we've added 85 post objects to our graph.

Figure 5-1 shows a 3D representation of the graph generated from Listing 5-4.

NOTE *The code to generate Figure 5-1 is in the 4th cell of the* Mastodon_network.ipynb *notebook.*

Each dot is a node representing a different user on the network, and each dashed line is an edge representing a post interaction between two users. Posts without replies don't create edges in the graph, and so aren't visualized here. Even though there's a lot of data and the graph looks like a mess at first glance, there are some takeaways. For example, you can already see this is a highly connected network. Looking at the nodes around the periphery, you can see that most users have a lot of edges leading to various other users, which means at some point they interacted through a post. Also note that some of the nodes have a lot more edges than others. Just as we did with the computer network in the previous chapter, let's begin to untangle this cloud of connectivity to see if we can make any interesting observations related to our research.

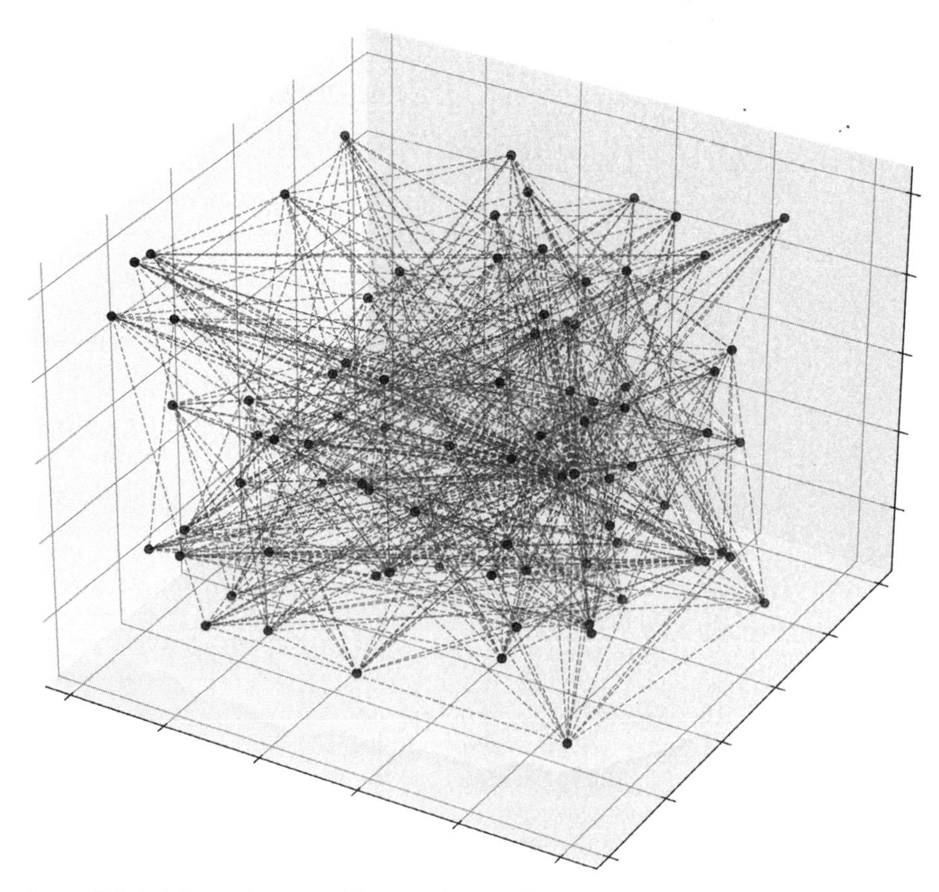

Figure 5-1: A 3D visualization of the social network graph

Network Analysis Insights

With our graph built, we can turn our attention to our research questions, starting with how much information gets propagated within the network.

Calculating Information Propagation

Calculating the amount of information something contains is an age-old problem with a lot of deep mathematical research behind it. Most truly useful methods deal with a concept called *information entropy* and dive into measuring the probability of some value (such as a phrase) existing by random occurrence. These measures are often complex to describe mathematically and would require a whole other discussion around linguistics and Markov chains. Instead, I've opted for the crude substitute of text length. Essentially, each post is treated as one unit of data and the post's total information is exchanged with each interaction. We consider information propagated when a user replies to a post.

We'll use a different method for the information exchange rate than the one we used in NetworkX (see "Examining How Machines Interact on the Network" in Chapter 4). Instead, we'll consider the *residual information (RI)* score, the difference between the amount of information added to a network and the amount consumed from it. For example, you could consider the residual information of your local library as the difference between all the books it has and all the books people in the area have read. It's very likely that there are some esoteric volumes that sit idle, waiting for the day someone will need them. The same can be said of a social network like Mastodon. When a user creates a new post, they add *potential information* to the network—that is, information waiting to be discovered by other users. When a user replies to an existing post, potential information converts to *kinetic information* through information exchange: information transfers from one user to another through the act of reading and responding to the original message. In this case, information flows from the origin user to the terminal user via a directed edge e, so the edge set p can also be viewed as the set of kinetic exchanges.

You can then reframe the question "How much information gets propagated?" as "How much potential information is required before some is likely to become kinetic?" or simply, "How many posts does it take before someone else is likely to read and respond?" One way we can answer this question is by calculating the ratio of original posts without replies (o) to original posts with replies (p); this gives us the RI score. For now, let's say a whole post is one unit of information, so, for each edge, one unit of information is exchanged from u to v. This is a *balanced exchange*, where all (and only) the information in the post is passed. If the node receiving the information could receive only half of it at a time, it would be an *unbalanced exchange* because the sender can send more than the receiver can handle.

Using these definitions, you can look at the overall tendency for information to spread through the entire network by comparing the ratio of potential information to kinetic information exchanged. The formula $RI = |p| \,/\, |o|$ describes the amount of potential information left in the network after all the exchanges have occurred. The result tells you approximately how much information must be added to the network before some of it is likely to be consumed by another user (by reading and replying). If every post on the network were responded to, you'd get an RI score of $0 \,/\, n = 0$. When there is zero residual information on the network, you need to add one piece of information for it to be consumed by another user. A network with no replies (all original messages) has an RI score of $n \,/\, 0$ = NaN, which indicates there is *only* residual information in the network, meaning no known amount of potential information will become kinetic. If there are twice as many original posts as there are response posts, the ratio is 2:1—for every two posts created, one post would get a reply. In another network with an RI score of 6 (a 6:1 ratio), only one in six posts get a reply, meaning the information flow is more resistant to propagating.

Listing 5-5 calculates the example network's RI score using the `DataFrame` object `post_df`.

```
❶ o_posts = post_df[post_df["in_reply_to_screen_name"].isna() == True]
  r_posts = post_df[post_df["in_reply_to_id"].isna() == False]
  if len(r_posts.index.to_list()) != 0:
    ❷ replied_to = r_posts["in_reply_to_id"].values
    ❸ o_no_r = o_posts.loc[o_posts["id"].isin(replied_to) == False]
      p_len = float(len(o_no_r.index.to_list()))
      o_len = float(len(o_posts.index.to_list()) - p_len)
    ❹ info_exchange = float(p_len / o_len)
  else:
      info_exchange = -1
  print("The RI score is: %.4f " % info_exchange)
```

Listing 5-5: Applying the RI algorithm to the example data

To measure the amount of potential and kinetic information in the network, we collect original posts (those that don't have a value for `in_reply_to_id`) into `o_posts` ❶ and reblogs into `r_posts`. We separate original posts that didn't receive any replies (p) into `o_no_r` ❸, and those that did (o) into `o_posts`, by gathering the IDs of posts with responses ❷ from the list of replies and creating a new list that excludes `replied_to` posts. The posts in `o_no_r` represent the potential information remaining in the network after the exchanges have all occurred. Finally, we take the ratio of the lengths of `o_no_r` and `o_posts` to get the RI score ❹. The result should be about `2.6358` for the sample data, indicating slightly fewer than three original posts are created for every one reply.

Identifying Cliques and Most Influential Users

A key aspect of network analysis is detecting smaller communities, or cliques, nested inside the larger network. Recall from Chapter 3 that a clique is a group of nodes that are all directly connected to one another. In the case of our social network, this would represent a group of users who are all familiar with each other and have interacted previously.

Let's find some cliques, starting by cleaning up the data set and displaying the graph. Cliques are meaningful only for nodes with connections to other nodes, so first we need to clean up the data to include only those posts with replies. We can drop posts without replies from the `DataFrame` like so:

```
r_posts = post_df[post_df["in_reply_to_id"].isna() == False]
```

All posts whose `in_reply_to_id` field is populated will be grouped in the `r_posts` object. We already discussed the second research question, "What cliques exist in this network?" from a theoretical perspective in Chapter 3, so let's apply that knowledge to understand the underlying structure of this network. A clique is a subset of nodes u wherein all the nodes of u are directly connected to one another, so if we assume that users read replies to their posts, we can defensibly loosen our directed graph to an undirected graph for the purposes of identifying these cliques. Listing 5-6 converts the graph and then finds the cliques as a list. We'll continue our analysis using the directed graph in combination with the clique list from the undirected graph.

```
uG = nx.to_undirected(G)
cliques = list(nx.algorithms.clique.find_cliques(uG))
```

Listing 5-6: Converting to an undirected graph to find cliques

Cliques in the network are interesting because they provide a picture of which users interact. Larger cliques usually represent users with some common association; they can reveal the formation of alliances and even predict fractures. Cliques by themselves may also be interesting: they tell you who knows whom, for example. However, it's when you start to analyze the members of different cliques that you really gain insight. You might identify the leaders of the cliques to see who has influence or status over the rest of the network. That's exactly what we'll do: we'll take what we've learned about the underlying cliques in the network to find which groups contain the most influential users in our Mastodon-like network.

The out-degree of a node in this case indicates the number of times other users have replied to a post the original node authored. A node with a high out-degree could be viewed as "more popular" since those posts tend to trigger more responses. By identifying the nodes who are close to this popular node, we can zero in on the underlying influencer. Listing 5-7 finds the node with the highest out-degree in the directed graph, then finds the cliques containing this node.

```
❶ deg_ct = G.out_degree()
  sorted_deg = sorted(deg_ct, key=lambda kv: kv[1])
  top_source = sorted_deg[-1]
❷ source_cliques = [c for c in cliques if top_source[0] in c]
❸ sG = G.subgraph(source_cliques[0])
```

Listing 5-7: Finding all maximal cliques for the highest out-degree node

First, we get the out-degree for all the nodes in the directed graph ❶. When analyzing relationships, you may sometimes want to quantify the strength of the connection between nodes along with the rest of the data. For example, if you know two users in the network are married, you may want to weight the edges between them higher than an edge between two people who are coworkers. In Listing 5-3, we captured the length of the text as a crude measure of the amount of data exchanged. We can use this information now to rate the quality of communications between users. To account for the quality of edges as well as the number, we replace the simple out-degree measure with a weighted out-degree measure like Dijkstra's algorithm (as I mentioned in Chapter 4, you do so by explicitly passing the `weight` parameter to the shortest path algorithm). After sorting the nodes in ascending order by out-degree count, we select the last item, the user who is the top source of posts that get responses, as the target node, and then use a list comprehension ❷ to extract the cliques that contain the target node.

Figure 5-2 shows the subgraph created by selecting the first of these cliques ❸.

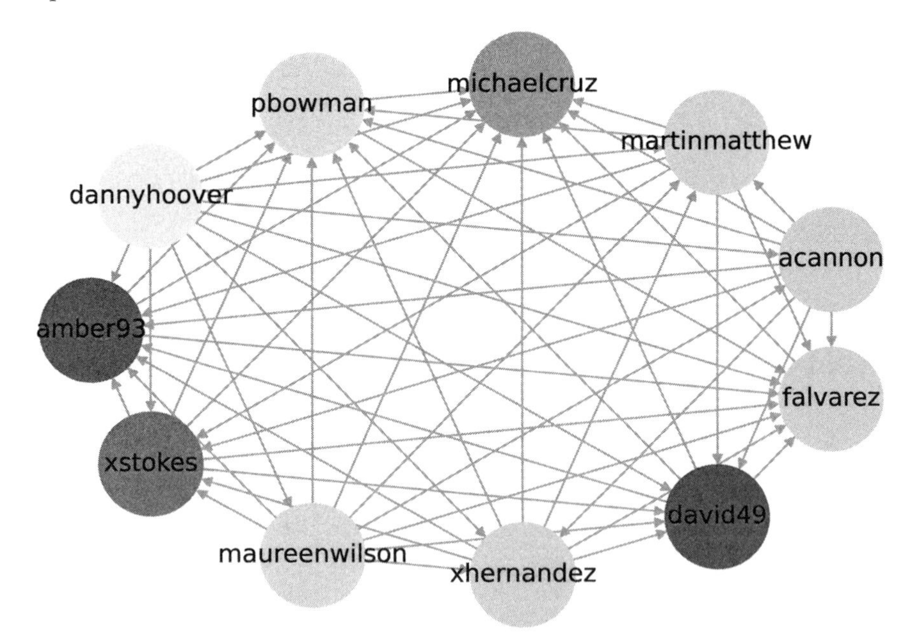

Figure 5-2: The clique subgraph for the user most responded to

NOTE *The code to generate Figure 5-2 is in the 13th cell of the* Mastodon_network.ipynb *notebook.*

The popular user, `dannyhoover`, has an outbound edge to each node in the graph. The users `michaelcruz` and `falvarez` reply to the largest number of other clique members. You can infer that `dannyhoover` is likely to be more influential (for these clique members) than either `michaelcruz` or `falvarez`. That isn't to say that those two users aren't influential in other contexts. Remember, when working with subgraphs, the information you derive is always with regard to the subgraph, not the graph as a whole.

For the third research question, "Who are the three most influential users?" we just extend the code in Listing 5-7 to consider the top three source nodes. Influential users are those who add potential information that's more likely to initiate a kinetic exchange. As an exercise, try to determine if the top three influential nodes are in the same clique. What can you possibly infer from the result?

Finding the Most Influenced Users

The next question posed explores the inverse relationship in the graph; that is, "Who are the three most influenced users?" is related to nodes with the highest in-degree. If you consider our definition of influence for this network, an influencer is someone who creates an original post that's likely to get a response from one or more users. In contrast, a highly *influenced* user is one who responds to a lot of other users' original posts. Luckily, the code is very similar to Listing 5-7. Simply swapping `G.in_degree` for `G.out_degree` produces a graph similar to the one in Figure 5-3.

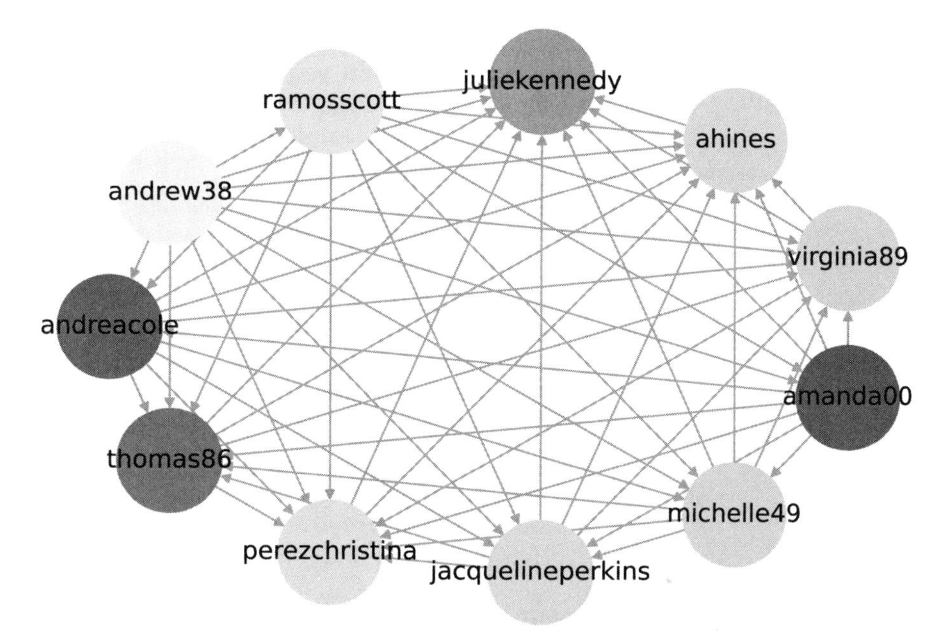

Figure 5-3: Finding the most influenced user

NOTE *The code to generate Figure 5-3 is in the 16th cell of the* Mastodon_network.ipynb *notebook.*

The user `juliekennedy` is responsible for a large portion of the kinetic information exchange in the network, which means they reply to the most users. Given our assumption that the person responding to a post has been influenced somewhat (at least enough to create a response), we can conclude that the user `juliekennedy` has been influenced by the largest number of users. Of course, you're free to (and probably should) debate the validity of this assumption. We're dealing with an area of security where you must be prepared to defend the assumptions you build into your analysis. When analyzing something as complex as human interaction, keep in mind there are limits to the accuracy and validity of the claims we can make.

Using Topic-Based Information Exchange

Going a little off-track on our research questions, we can answer the two previous questions of influence for more specific post topics using *topic-based information exchange*, in which we consider the most influential and influenced users within a certain context or topic. For instance, we might consider the most influential heart surgeon or the most influential hacker. By examining influence and popularity with a contextual example, we can gain more insight into the interactions we've recorded. Simply put, we can answer "What are these user interactions about?" We'll find the most influential and influenced users for particular topics, such as environment and politics, but you can extend the same principle just as easily to search for users discussing current events or other topics of interest.

For topic-based information exchange, we use the *Hyperlink-Induced Topic Search*, or *HITS* (also known as *Hubs and Authorities*), an algorithm for analyzing the link relationships in a directed graph.[1] Originally designed for internet search engines to score web pages on their relevance to a given topic, HITS has been adapted to many other types of link analysis. In terms of security and social network analysis, HITS can give useful context to the concept of generic influence measure, like information exchange ratio (IER). For example, security researchers used Twitter to track information related to a terrorist attack in Mumbai[2] by examining topics related to the attack and determining which users seemed to have the most authoritative understanding of the events.

The intuition behind the original algorithm is fairly simple: certain sites, known as *hubs*, serve as large website directories. Pages are sorted by relevance to a queried topic. A good hub is one that points to many other pages across many subjects. If multiple hubs point to the same source page for a topic, that page is considered to be an authority on the subject. In other words, an authoritative node represents one that is linked to by many different hubs. The higher the hub scores, the more authoritative the node. The more authoritative nodes a hub connects to, the higher its hub score becomes. Modern search engines are excellent examples of hubs. These sites aren't authoritative on any one topic they catalog, but they can lead users to other sites that *are* authoritative.

In our network, a hub would be a user whose post on a subject is reblogged by a large number of authoritative users. On the other side of the information flow are the authority nodes, which equate to users who reblog the information from several quality information hubs. NetworkX relies on the SciPy library under the hood to convert the graph into a *sparse adjacency matrix* (a list where every possible connection in the graph is recorded as either present in the data or not). In turn, SciPy relies on NumPy to handle the matrix math. Unfortunately, this dependency chain can be fragile. Depending on how you install the packages, you might get an attribute error like `module 'scipy.sparse' has no attribute 'coo_array'` when running the *Mastodon_network.ipynb* file. I was able to temporarily resolve this by installing NetworkX version 2.6.3 using the command:

```
conda install -y networkx=2.6.3
```

The HITS algorithm is performed iteratively over a subset of relevant nodes, typically returned from some search algorithm. With each iteration, the algorithm recalculates the two real values representing the hub score and authority score for each node in the subset. Since the hub score for a good hub should increase with each new iteration, the score it lends to each authority will also increase, and vice versa. The final output is two scores for every node in the subset.

Listing 5-8 shows a method to find hubs and authorities relating to posts containing the word *environment*, using the DataFrame object from Listing 5-2 once again.

```
❶ post_df["content"] = post_df["content"].str.lower()
❷ env = post_df[post_df["content"].str.contains("environment")]
❸ repl = post_df[post_df["in_reply_to_id"].isin(env["id"].values)]
  hG = nx.DiGraph()
  for idx in repl.index:
      row = repl.loc[idx]
    ❹ hG.add_edge(row["in_reply_to_screen_name"], row["user_screen_name"])
❺ hub_scores, auth_scores = nx.hits(hG, max_iter=1000, tol=0.01)
```

Listing 5-8: Building a topic-based subgraph and running the HITS algorithm

We begin by casting the post text to lowercase ❶ (so we can perform case-insensitive matching), and then use the built-in pandas contains function for locating rows based on text content to retrieve all posts with the related root word ❷. This will also match environment*al*, environment*alist*, and so on. We use each row's post ID to extract the set of responses to these posts of interest ❸. We loop over each of the reply rows and create a directed edge, which indicates the flow of influence for the related topic, in the resulting subgraph ❹.

Finally, we use the resulting subgraph to calculate the HITS hub and authority scores ❺. The max_iter parameter passed to the networkx.hits function (part of the NetworkX core library) controls the maximum value of iterations the algorithm will run in cases where the code doesn't converge on a solution (see the NetworkX documentation for a description of how the HITS algorithm reaches convergence). The tol parameter controls the error tolerance to check for convergence. If the algorithm fails to converge on an answer within the tolerance and max iterations, a PowerIterationFailedConvergence exception will be raised.

The algorithm starts from the assumption that all nodes have a hub score and authority score of 1. At each subsequent step, it computes two update rules:

Update authority scores

Update each node's authority score to be equal to the sum of the hub scores of each node that points to it. That is, a node is given a higher authority score by reblogging messages of users recognized as information hubs. This is represented by:

$$auth_{(topic)}(u) = \sum_{i=0}^{n} hub_{(topic)}(v)$$

where n is the number of incoming references to u, and v is the node at the opposite end of the ith edge.

Update hub scores

Update each node's hub score to be equal to the sum of the authority scores of each node that it points to. In our example, a node is given a high hub score by writing posts that are reblogged by nodes considered to be authorities on the subject. This is represented by:

$$hub_{(topic)}(u) = \sum_{i=0}^{n} auth_{(topic)}(v)$$

where n is the number of outgoing references from u, and v is the node at the opposite end of the ith edge.

You can now reframe the second and third research questions in terms of a given subject. For instance, "Who are the top three hubs for the topic of environment?" and "Who are the top three authorities for the topic of politics?" The topic-based subgraphs in Figure 5-4 show the results from our sample data.

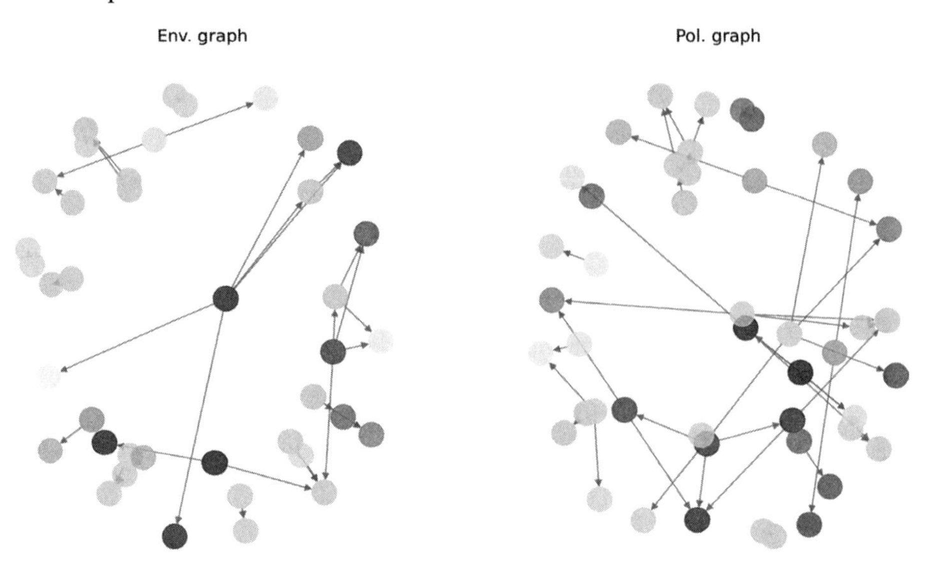

Figure 5-4: Topic subgraph examples for environment and politics

NOTE *The code to generate Figure 5-4 is in the 32nd cell of the* Mastodon_network.ipynb *notebook.*

Rather than labeling the nodes, I used the `nx.spring_layout` function to visually graph the influence structure for the two topics. According to the documentation,

> The [spring layout] algorithm simulates a force-directed representation of the network treating edges as springs holding nodes close, while treating nodes as repelling objects, sometimes called an anti-gravity force. Simulation continues until the positions are close to an equilibrium.

This has the effect of pushing highly connected nodes more toward the center depending on the relative connectivity of the rest of the nodes. Nodes near the center have exerted influence on more users, so the other nodes have pushed farther away. You can see that the graph for politics on the right of Figure 5-4 has a larger number of small clustered influences near the edges of the graph, with only a few nodes showing more influence than the others. The environment graph on the left, however, shows a distinctly influential user near the center and then a few smaller clusters of local influence around the edges. When using the `spring_layout` function, keep in mind that the initial positions are randomized so the resulting graph is stochastic (random). Rerunning the code will likely result in a different visual layout, but the most influential nodes will always have pushed the other nodes farther away than less influential nodes.

After running the HITS algorithm, you should find that the top three hubs for environment (in descending order of hub score) are `williamclarke`, `victoria73`, and `nromero`. The top three authorities for politics (also in descending order of authority score) are `wernerbrianna`, `trivera`, and `susanjohnson`. Remember that the scores produced by the HITS algorithm are relevant only to the topic subset. A node with a high authority score for "pet food" wouldn't score the same on "programming."

At the start of the chapter I mentioned how social network connections and influence could be used to predict the spread of malicious content; this is your first real method for doing so. A lot of malware is spread through social network message attachments. Once you identify a malicious message on your network (and extract some useful topic information), you can leverage the HITS algorithm to predict which users are more likely to respond to the message. By doing so, you can deal with the risks in descending order of importance. A real-world example of this occurred as I was revising this chapter. During the height of the COVID-19 pandemic fear, attackers used an infected version of a tracking map to trick concerned users into visiting a malicious website. Once this story broke (*https://krebsonsecurity.com/2020/03/live-coronavirus-map-used-to-spread-malware*), security teams used the HITS algorithm to track which, if any, of their users might have been impacted.

Analyzing Network Organization

The final question we want to answer—"Who could introduce the most new connections?"—is a bit more complex but important nonetheless. Researchers and analysts use this type of information when analyzing the organization of networks from street gangs to military battalions—anywhere individuals may not directly interact but share some common oversight "higher up the ladder." For example, a soldier in unit A may send information about enemy troop movement to the unit commander, who in turn forwards the information to the base commander. The base commander is in communication with several different unit commanders at any given time and may send the message to another unit commander in unit B, who then moves to intercept the enemy. In the US, this chain of command is an implementation of a node ancestry that can be traced from the office of the president (as commander in chief) all the way to each individual soldier in boot camp. By examining which nodes can facilitate connections between large numbers of currently disconnected nodes, you can begin to understand each person's importance in the hierarchy.

Figure 5-5 shows a tree structure for an example that's probably more familiar to you, a company organization chart.

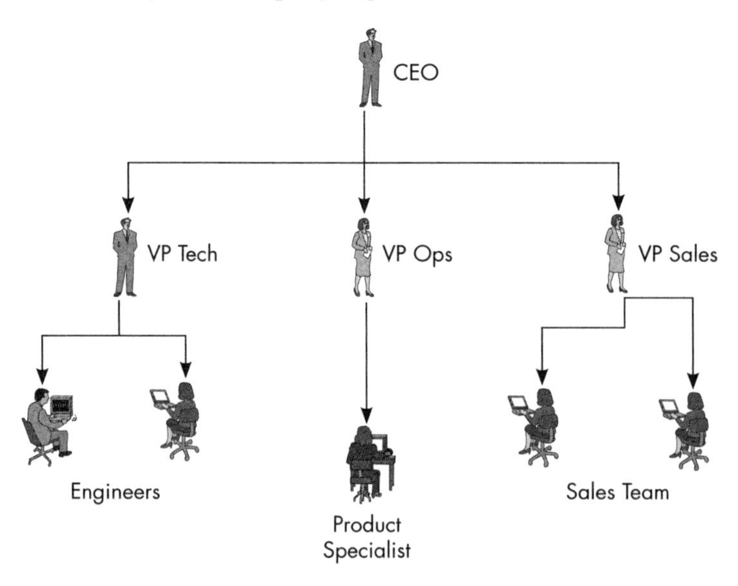

Figure 5-5: An example tree from an organization chart

The root of this tree is the CEO at the top, below whom are three managers who all report directly to him. Below each manager are the subordinates that make up their team. Understanding the influence within social structures is vital to planning (or circumventing) security controls intended for interaction with humans, such as social engineering; social engineers use the concept intuitively to gain legitimacy with other employees. Simply put, if you can convince an influential person to introduce you, you can bypass most resistance. Of course, you wouldn't want to directly call the CEO of a large company if the branch manager is capable of making the introduction you need. The first common node between yourself and the person you'd like to be introduced to is the *lowest common ancestor (LCA)*.

To determine the LCA, we first need to define *node ancestry* as it relates to trees (rather than genealogy). In graph theory, a tree is a special type of graph structure in which any two nodes are connected by exactly one path (*https://mathworld.wolfram.com/Tree.html*). The node at the start of the tree is the root node; offspring nodes are called branch nodes unless an offspring branch has no branch nodes of its own (a dead end), in which case it's called a leaf node.

By definition, a simple graph has no directionality and no cycles. A *polytree* extends the concept of a simple graph to include directionality, making a *directed acyclic graph (DAG)*. This seemingly simple change imparts a lot of interesting properties. For example, a DAG has a *topological ordering*: the nodes are ordered so the root node has a lower value than the leaf nodes. DAGs are one of the most studied of all graph structures because they appear so frequently in nature. From the literal branching of trees and plants, the veins in your body, and rivers to the structure of most computer programs, DAGs can represent a huge number of natural and artificial systems. In our case, using a DAG to represent the relationship between nodes will allow us to encode a hierarchy of membership in the social network.

Node ancestry for arbitrary polytrees is similar in concept and structure to that of a family tree. However, the order relies on the topological sorting of DAGs, rather than being strictly chronological. The most influential users are those nodes with some amount of out-degree and no in-degree (users, like `dannyhoover`, who have more influence than other users) and these form the roots for distinct trees. Each node they influence becomes a branch in the tree. For each branch node, the out-degree edges are again added as branches. Branching continues until all nodes are placed. This leads to leaf nodes with an out-degree of zero and some amount of in-degree (the users most influenced by other members of the network). Ordering the nodes in this manner gives you an idea of the flow of influence.

Formally speaking, an ancestor of node u is any other node v such that a directed path exists in the graph from v to u or, written more algebraically:

$$Ancestry(u) = (u \leftarrow v) \in E$$

The *common ancestors* for two nodes (u,v) are any nodes x that have a directed path to both u and v in the set of edges. This can be written as the intersection of these two subsets of edges:

$$CommAnc_{(u \wedge v)} = (x \rightarrow u \in E) \cup (x \rightarrow v \in E)$$

The LCA of two nodes (u,v) is the common ancestor with the shortest path distance to both nodes, which is also the ancestor with the maximum path length from the root of the graph. For example, you and your cousin share some of the same great-grandparents. However, you also share some of the same grandparents. While both your great-grandparents and your grandparents are your ancestors, since your grandparents are closer to your generation than your great-grandparents are, they would be your LCA. Figure 5-6 shows two examples of ancestry on the same tree.

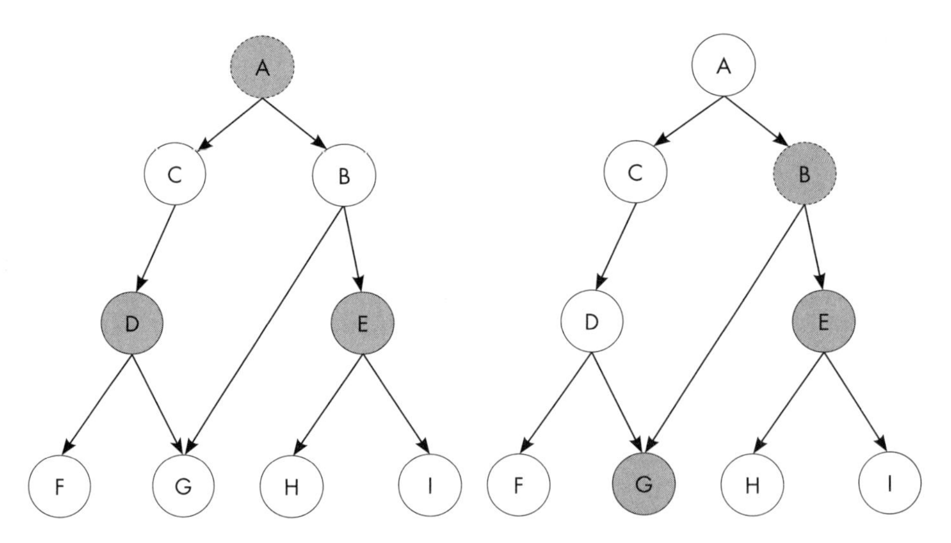

Figure 5-6: A general ancestry illustration

In each tree the shaded node with the dashed outline is the LCA of the other two shaded nodes. On the left, the LCA for the nodes *D* and *E* is the root of the tree, *A*. On the right, although node *A* is still a common ancestor, node *B* is farther from the root and therefore the LCA. Thinking about this in terms of information security, the LCA of two nodes is the closest potential pivot point between them. If a user at node *G* wanted to be introduced to the user at node *E*, they could ask the user at node *B* to introduce them. In Listing 5-9 we tally the occurrence of each node as the LCA of other node pairs.

```
❶ ancestors = list(nx.all_pairs_lowest_common_ancestor(G))
pred_count = {}
for p, lca in ancestors:
  ❷ if p not in G.edges():
        if lca in pred_count.keys():
          ❸ pred_count[lca] += 1
        else:
          ❹ pred_count[lca] = 1
sorted_pred = sorted(pred_count.items(), key=lambda kv: kv[1], reverse=True)

for k in sorted_pred[0:5]:
    print("%s can bridge %d new connection" % (k[0], k[1]))
```

Listing 5-9: Counting LCA occurrences for all nodes

First, we generate the list of ancestors using the NetworkX function `nx.all_pairs_lowest_common_ancestor` ❶, which returns a dictionary where the key is a pair of nodes from the graph, and the value is the LCA node for that pair of nodes. With the `ancestors` list populated, we then use a `for` loop to assign the pair of nodes to the variable `p` and the resulting ancestor to the `lca` variable, in order to count how many connections `lca` can bridge.

We ignore pairs of nodes with an edge between them, since one of the nodes is the direct ancestor of the other ❷. For example, the pair of nodes *B* and *E* from Figure 5-6 can be ignored, even though the NetworkX function produces the LCA of the pair. For each pair of nodes without a direct edge between them, we check if their `lca` is in the `pred_count` dictionary, which counts the number of times a node is the LCA for two other nodes. If the LCA node is already in the dictionary, increment the count by 1 ❸. Otherwise, create a new entry with a value of `1` ❹. Running this code will print the top five users along with the number of connections they can potentially bridge, as shown in Listing 5-10.

```
georgejohnson can bridge 444 new connection
dannyhoover can bridge 444 new connection
vkhan can bridge 372 new connection
judith20 can bridge 336 new connection
david49 can bridge 216 new connection
```

Listing 5-10: The results of the LCA analysis

NOTE *You can see how I calculated these results in the 27th cell of the* Mastodon_network .ipynb *notebook.*

The root user, `dannyhoover`, is tied for first and can potentially bridge 444 new connections within the network. Since we already think this user is very influential, that may come as no surprise. Their position at the root of the tree also means they're the last possible LCA for all pairs of nodes, if no other ancestor can be found, so this result may not be as interesting as the second and third place. The fact that the user `georgejohnson` got the same exact score as `dannyhoover` is interesting and may point to two structures in the data worth investigating.

The user `judith20` can bridge 336 connections. As an exercise, examine how this user fits into the structure of the tree. Who influences their activity (inbound edges) and who do they influence (outbound edges)? What measures of centrality do they score most highly on?

The Proof of Concept: Social Network Analysis

The proof-of-concept code for this chapter, found in the *social_network/post_graph.py* file in the book's resources, allows you to capture the post data from your personal timeline into JSON data you can analyze using the methods shown here.

You'll need to register for an account on whichever Mastodon instance you choose (I'm using defcon.social). You'll then need to register an application for your own set of API credentials (*https://docs.joinmastodon.org/client/token*). Registering an application gets you an API token and API token secret that identifies a specific application under your account and grants access to authorized functions (such as liking posts and following users). Depending on the Mastodon instance you choose, you may be required to answer some questions to qualify for different use cases; otherwise, you simply need to define the scope of the access token as you create it. A lot of Mastodon instances are friendly to researchers, as long as you plan to protect the privacy of individuals' data.

You'll be given a unique API key that identifies your API account to the Mastodon instance, paired with an API secret that should be protected like other cryptographic keys.

Once you've registered, you can use the API via the Python Mastodon library to scrape your own timelines. Refer to the Mastodon library documentation (*https://mastodonpy.readthedocs.io/en/stable*) and the Mastodon API documentation (*https://docs.joinmastodon.org/api*) to get a sense of the data that's available and how to access it using this library.

Listing 5-11 shows the proof-of-concept code.

```
❶ from mastodon import Mastodon
  import pandas as pd

❷ ACCESS_TOKEN = "YOUR-TOKEN-HERE"
  BASE_URL = "https://defcon.social"
❸ m = Mastodon(access_token=ACCESS_TOKEN, api_base_url=BASE_URL)

❹ timeline_data = m.timeline(timeline="public")

  df = pd.DataFrame(timeline_data)
  df["id"] = df["id"].astype(dtype=str)
  df["in_reply_to_id"] = df["in_reply_to_id"].astype(dtype=str)
  df["in_reply_to_account_id"] = df["in_reply_to_account_id"].astype(dtype=str)

  print(df.info())
❺ df.to_csv("mastodon_timeline.csv")
```

Listing 5-11: Capturing Mastodon public timeline data to a CSV file

First, we import the mastodon library ❶. Once we have the API credentials, we modify the template file with the access token and base URL ❷ and run it from our terminal. The code uses these credentials to create an authenticated API object ❸, used to retrieve the timeline data ❹, which is conveniently delivered as a dictionary suitable for JSON encoding. We loop over these results and write them into the output CSV file ❺.

NOTE *You can pass a number of parameters to the* timeline *function to change the amount of information contained within each JSON object. See the library documentation for the function at* https://mastodonpy.readthedocs.io/en/stable/07_timelines.html. *Experiment with these options to find the right balance of data for your analysis.*

Now you can use code similar to Listings 5-2 and 5-4 to read the data back into a pandas DataFrame, then mold it into significant features and finally a relevant directed (or undirected) graph using NetworkX. You can also bypass writing to an intermediate file by combining this code with a data processing pipeline to analyze status information in near real time. We'll discuss processing pipelines in Part III.

The Darker Side of Social Network Analysis

Hopefully now you have an idea of how quickly and easily someone can build a map of someone else's social life. The important thing to remember is, like maps, social network graphs require interpretation. When we interpret social network information, we're invariably viewing the data through our own social biases. The core of the issue is that we're trying to reduce a highly complex, multifaceted problem, such as the motivation behind people's interactions, into a tightly controlled and well-defined mathematical model. To do so, we have to apply heuristics we choose based on our own social experience. For example, I mentioned earlier that you might want to weight interactions between married couples higher than those between coworkers. This shows one of my heuristic biases, which comes from my experiences, education, and understanding but doesn't necessarily reflect the reality of everyone's situation. You'll need to make many such assumptions when building a SNA model, and it's important to understand when, where, and how much you're allowing your own biases to impact the analysis. This is one of the primary reasons I recommend doing SNA with a team. Peer review, especially from peers of diverse backgrounds, is one of the best counters to the problems inherited from a single-perspective interpretation.

The other reason I recommend caution is that SNA raises moral and ethical questions. It is perhaps the dark arts of applied mathematics in security, primarily because there can be very real and dangerous consequences when it is abused. SNA has been used by tyrannical governments to attack dissidents, threaten whistleblowers, and manipulate the people of a society. Unfortunately, not all ethically questionable uses of SNA are as easy to spot. There are tools and sites designed to make it even easier to collect someone's publicly available (and sometimes private) information. We live in a world that constantly struggles to balance privacy and openness. The small-world experiment can be used to link movies to Kevin Bacon or to link each one of us to any number of criminal figures and organizations. As an analyst, you're responsible for understanding what's ethically and morally appropriate.

Summary

Although this chapter demonstrates the concept of social network analysis using Mastodon as an example, none of these concepts are inherently tied to the Mastodon platform. The US government and university researchers have been working on different technologies to leverage the information obtained from analyzing the reply network structure of discussions in dark web forums to understand the extent to which dark web information can be useful for predicting real-world cyberattacks.[3] In his paper "Tracking, Destabilizing and Disrupting Dark Networks with Social Networks Analysis," written for the US Navy, Sean Everton covers SNA as a means to develop strategies for tracking and disrupting criminal and terrorist networks.[4] The paper serves as both a tactical and strategic introduction, so I highly recommend reading it.

As you extend your own SNA to work in the wild, you'll want to reference the API documentation for whatever social network you're using. If no API exists (or the platform starts charging exorbitant rates), you may have to resort to old-fashioned web-scraping techniques to gather the data you need. Such tasks are outside the scope of this book, but there are many excellent materials for doing so.

So far, everything we've used graph analysis for has been focused on the past. You can think of this as *descriptive* security analysis because it aims to classify things as they are now (or as they were when the data was captured). *Preventative* security analysis, however, seeks to analyze what might occur in the future so that hopefully we can step in and prevent a security incident from occurring in the first place. To achieve this, we'll use one of my favorite simulation algorithms, Monte Carlo simulations—the subject of the next chapter.

6

ANALYZING SOCIAL NETWORKS TO PREVENT SECURITY INCIDENTS

In the last three chapters on graph theory, we've built graphs from a snapshot of a network at a particular moment in time; that is, we've worked from fixed, historical data. But finding and responding to events in the past always leaves the white hats one step behind the black hats. If we want to know more about what happened before or after the time captured in the data, we need new analytic techniques. The future requires *predictive analytics*, a branch of mathematics that aims to statistically determine the probability of future or past events given some set of known observations. The goal is to stop the security incident before it ever gets started. To achieve this, though, we need a way to predict how things will change over time. We'll use a specific algorithm, the Monte Carlo simulation, to model network activity that hasn't occurred yet. While this chapter presents the topic in the context of social network analysis, Monte Carlo simulations are suited to a wide variety of topics and network types. For example, I've used Monte Carlo simulations to predict which machine an adversary would attack next.

Here, we'll attempt to predict the answers to the following questions about a social network:

- How far is information likely to spread from a given node?
- Which nodes are being influenced by other nodes?
- What links could be severed to disrupt the flow of information between two nodes?

From a security perspective, these questions assess the resilience of a social network in the face of adversarial behavior. They ask, "How easy would it be to break up an association of people?" Companies ask these questions about themselves to determine if they could withstand losing key employees, facilities, or vendors. Law enforcement asks them when they assess a criminal syndicate.[1] Criminals also ask these questions about an organization when they want to select the targets for spear phishing and other social engineering attacks.[2]

We'll begin this chapter by looking at how to define and construct a Monte Carlo simulation. We'll discuss how different levels of randomness can be applied to replace unknowns. Then we'll use the Monte Carlo simulation we've built to predict the way a piece of information might move through the social network from Chapter 5, given previous observations. Finally, in the proof of concept for this chapter, we'll see how to modify our simulation to account for adversarial behavior. By the end of this chapter, you'll be able to use your knowledge of graph theory and apply Monte Carlo simulations to predict the outcome of different scenarios on your own social networks.

Using Monte Carlo Simulations to Predict Attacks

For the rest of this chapter to make sense, we need a little bit more theory on top of the graph theory we've already covered. Specifically, I've been throwing around the word *simulation* without really defining it. Generally speaking, a simulation is a controlled imitation of a real-world process. Simulations rely on models to describe the key characteristics and behaviors present in the simulated environment. The simulation code acts as the manager of the model, choosing various actions and applying them to evolve the model at each step. Modern models and simulations are most often designed using a combination of programming languages like C and Python, where C is used for critical functions and user-friendly Python syntax is used for the rest. Luckily, all the underlying C code has already been handled for us, so we can focus on the Python interface.

In theory, any phenomenon that can be reduced to data and equations can be simulated on a computer. In practice, however, simulation is difficult because most real-world processes are subject to a practically infinite number of influences, and it's impossible to account for them all.

A Monte Carlo simulation is a way of quickly gathering statistics about some seemingly random (or at least hard to predict) variable, given a set of constraints. Unlike other forecasting methods, which work with a set of fixed input values, a Monte Carlo simulation predicts a set of outcomes based on an estimated range of values. You've probably seen the results of a Monte Carlo simulation in the form of a storm path map (sometimes called a *spaghetti model*). Monte Carlo simulations are most useful when the probability of varying outcomes can't be determined because of random variable interference. A Monte Carlo simulation focuses on repeating the test with random samples to achieve certain results. It also helps to explain the impact of risk and uncertainty in prediction and forecasting models because the values for the random variables are chosen using the distribution of previously recorded values. The larger the variance in the random value, the more variance in the different results of the simulation. In principle, Monte Carlo methods can be used to investigate any problem with a probabilistic interpretation.

In a security context, I've used Monte Carlo simulations to predict and interrupt attacks. To do so, I programmed some rules that mimicked the previous decisions of the attacker and ran thousands of simulations to predict where the attacker would end up. My team created a network graph (similar to the one from the previous chapter) in which we weighed the ease of access along with the machine's attractiveness to the attacker (in terms of data or lateral movement). We then ran simulations with the attacker starting from random machines we knew had been exploited and using a stochastic process to determine if the attacker could successfully move from one machine to another.

We had additional rules to define how the attacker selected machines and so forth, but the question we were trying to answer was simple: After six days of active exploitation, which machines had the highest probability of being infected? In math terms, the law of large numbers tells us that integrals described by the expected value of some random variable can be approximated by taking the empirical mean (sometimes called the *sample mean*) of independent samples of the variable. In lay terms, the machines with the highest probability in our network simulation tests were likely those with the actual highest probability. And there's our definition for "predicting" the future: we can state, with some degree of confidence, the statistical probability of each outcome. Unfortunately, that means things won't always turn out as predicted.

Modeling changes requires that we first have a way to describe what can and can't happen. We'll use a mathematical construct known as a finite state machine to handle this task. We then need to create a fake world for our simulation to inhabit. NetworkX will fill this role by providing the graph of our social network. Finally, we need some way of recording the different events so that we can analyze them. This is where the Monte Carlo algorithm really starts to take shape. Let's start by defining each piece, and then we can tie it all together with some different simulations.

Finite State Machines

A *finite state machine* (*FSM* or simply *state machine*) is a hypothetical machine that can be in exactly one of a finite number of states at a given moment in time, where a *state* is a unique configuration of variables. If you think of a board with three switches on it, each possible switch configuration represents a possible state for the board. It's called a *finite* state machine because you can count the number of possible states. In the example switchboard, if each switch can be in one of two possible positions, there's a total of eight possible configurations, or states, the switchboard could be in. If you think of these switches like bits in binary, you could represent the values between 000 and 111, or 0 through 7 in base 10. The state machine can change from one state to another in response to some external *input*, or decision (such as flipping one of the switches on the board). Changes from one state to another are called *transitions*.

Formally, a state machine M is defined by a quintuple $M = (\Xi, S, S_0, \delta)$ comprising a finite number of possible inputs (Ξ, the *input alphabet*), a set of all possible states (S), an initialization state (S_0) where $S_0 \in S$, and finally the conditions for each valid transition between states, δ. We can represent a state machine as a directed graph wherein each node is a potential state of the machine, and each edge is the required input to transition from state u to state v. Figure 6-1 shows a simple FSM graph with five states and four transition inputs.

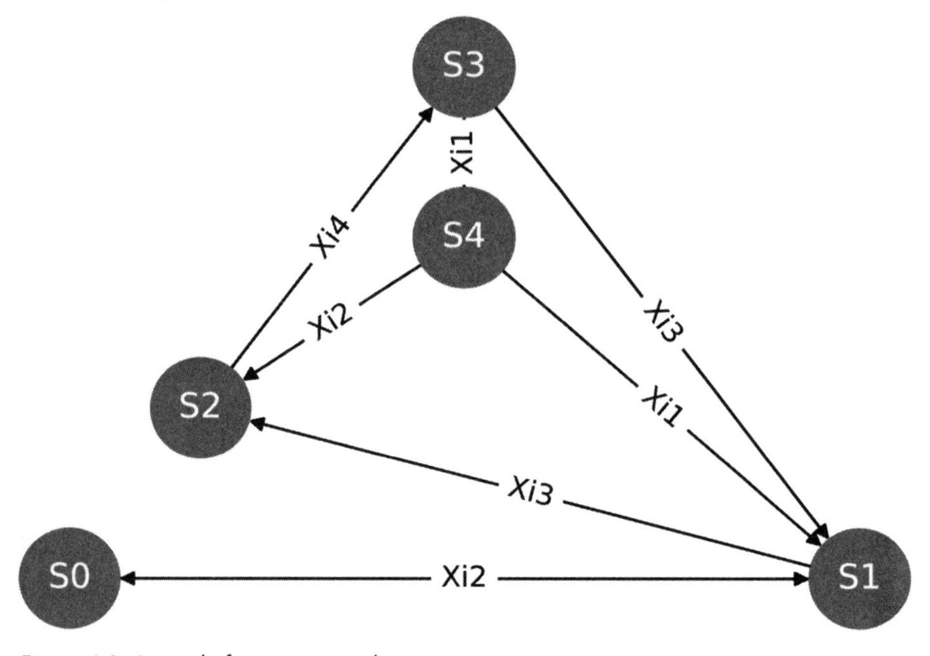

Figure 6-1: A simple finite state machine

NOTE *The code to generate Figure 6-1 is in the 1st cell of the* State_Machine_Graphs .ipynb *notebook.*

Looking at this graph, you might be confused; after all, I just said there were four transitions, but there are nine edges here (the bidirectional edges between S_0 and S_1 and S_3 and S_4 count as two each). This is because the same input may be used in multiple transitions. The inputs Ξ_3 and Ξ_2 in Figure 6-1 are both examples of this: Ξ_2 is used to transition between S_0 and S_1 as well as between S_4 and S_2, while Ξ_3 can be used to transition from

S_3 to S_1 or from S_1 to S_2. Think of the input Ξ_2 as an action, like flipping a particular switch. Depending on what state you're in currently, the action of flipping the switch may take you to a different state. If you're in S_0 and flip the switch, you end up in S_1. If you're in S_4 and flip the switch, you'll end up in S_2. The input hasn't changed—it's still Ξ_2—which illustrates an important relationship between inputs and states. The same input may result in arriving at a different state, depending on the current state.

FSMs are either *deterministic*, meaning each transition has a single guaranteed outcome, or *stochastic*, meaning the outcome of an input is influenced by randomness and not guaranteed to produce the same result every time. To illustrate the difference between the two types of FSMs, imagine picking up a pencil. In a deterministic world, attempting to pick up the pencil will always result in successfully picking up the pencil—or transitioning to the state where you have the pencil, in FSM parlance. In a stochastic world, you may fail to pick up the pencil with some probability $0 < p < 1$. If you fail to pick up the pencil, you transition to a different state than if you'd succeeded. Perhaps you dropped the pencil on the floor and you're now in that state instead. This is a very simplistic example, but the point is that stochastic FSMs allow randomness to influence the results. This is powerful for generalizing the description of complex interactions because you don't have to understand the mechanisms at work, you only need to measure the statistical distribution of possible outcomes and you can approximate the same phenomenon.

You'll often see a mix of deterministic and stochastic inputs in the same FSM. For example, in the FSM from Figure 6-1, Ξ_4 is deterministic. If you're at S_2, the input Ξ_4 is guaranteed to transition you to S_3 and there's no other possible outcome. On the other hand, Ξ_1 is stochastic: if you're at S_4 and select action Ξ_1, you might end up at S_1 or at S_3. If no probabilities are given for these outcomes, it's assumed to be uniformly random. If probabilities are given, the probability distribution is used in a weighted-random selection function. NetworkX has parameters for labeling the edges, which can be useful when showing the probabilities or, as I've done here, the transition names. You can see examples of this code in the accompanying Jupyter notebook. For more detailed examples of using FSMs, I highly recommend checking Wolfram Alpha.

NOTE *Other materials may denote the state machine's input alphabet as an uppercase sigma (Σ), which more commonly denotes the summation function.*

Now that you understand FSM structure a bit better, let's move on to how we can leverage it using an algorithmic gem known as random walks. Random walks allow us to repeatedly choose random inputs for our FSM to automate the simulation of these choices based on the rules we define.

Network Modeling with Random Walks

In math terms, a *random walk* is a series of randomly chosen steps (or transitions) within a system that result in a random final state after some number of steps. I like the analogy of a tourist wandering in an unfamiliar city. They may walk up the street a bit, decide to turn left, go a few blocks, and then decide to turn around and go back the other way. These walks are erratic and unpredictable by definition. Versions of the random walk model have been applied to research topics from economics to neurology, and now information security!

We're going to apply this methodology to model how people pass information to one another and ultimately utilize the network. We can then use this information to explore what might happen if we change some of the parameters (such as an attacker taking over one or more lines of communication) without risking actual disruption to the network. Randomly selecting a series of transitions within a state machine over n steps ($T(n)$) updates the state of the system with the result of the input. The subsequent decision must be based on the new state, and all actions may not be valid in all states. The set of valid transitions from a given state is denoted $\Xi_{(S)}$. At each step a transition is selected from $\Xi_{(S)}$ and appended to $T(n)$. We can write this as:

$$\frac{T_{(n)}}{rand\ (\Xi_{(S_0)})\ldots rand\ (\Xi_{(n)})}$$

The state is updated and the process is repeated until all n steps have been taken, or no valid state transitions are left. The resulting *terminal state* is the product of applying the random walk defined in $T(n)$ to the state machine M ($M \times T(n) = S(Tn)$).

As a concrete example, let's define a simple state machine. Imagine you're standing in the center of a large empty room. This is the initial state, S_0. On the floor is a 7×7 grid of squares, and positions in the room can be expressed as location tuples (x, y) on the Cartesian plane (your position is $S_0 = (4, 4)$). You can move forward, backward, left, or right one square with each step. Given an arbitrary set of instructions, you may end up standing on any square in the room; therefore, each square can be viewed as a potential state in S. The inputs [*forward, backward, left, right*] form the input alphabet Ξ. The two diagrams in Figure 6-2 show the same uniformly random walk for $n = 10$ in two dimensions on the left and three dimensions on the right.

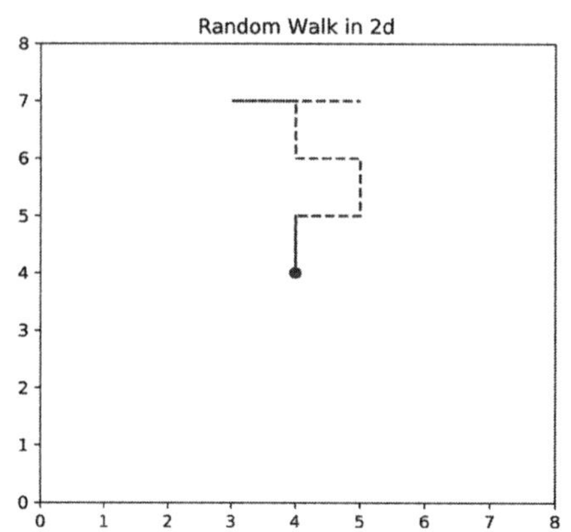

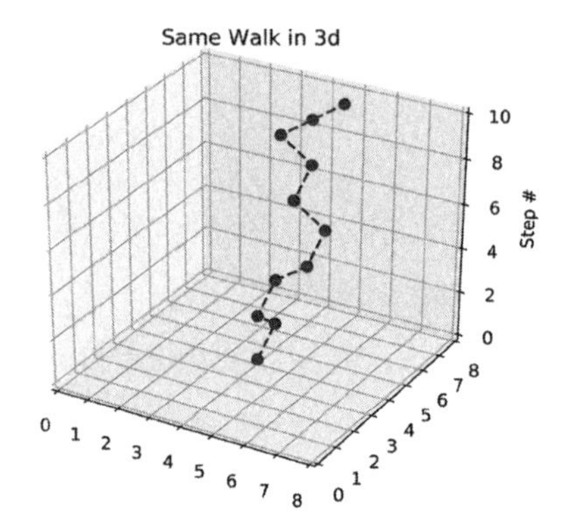

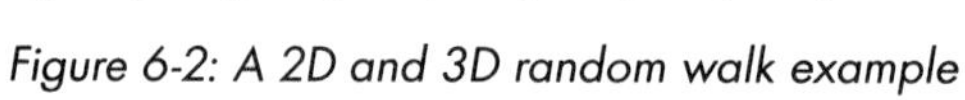
Figure 6-2: A 2D and 3D random walk example

In the 3D example, the third dimension is time (you wouldn't actually start to levitate as you moved around the room).

NOTE *The code to generate the two images in Figure 6-2 is in the 3rd cell of the* State_Machine_Graphs.ipynb *notebook.*

In contrast to a uniformly random walk, where each input is equally likely, in a *biased random walk* (or just *biased walk*), one or more of the inputs is likely to occur more than the rest. In a biased walk, we extend Ξ to a set of tuples: (*input, probability*).[3] At each step we select one of the inputs from the list using a weighted random selection function—that is, one that respects the probability distribution we pass to it. We'll construct a version of this later, but for now the key takeaway is that biased walks allow you to add any a priori information you have about behavior probabilities to your modeling. For example, if you know there's a malicious actor who's looking for financial information, you may choose to bias your model of their behavior toward nodes in the network that have access to such information.

Up until now we've covered what a state machine is and how we can use one to simulate a series of choices. Because a random walk represents a single set of choices made within a stochastic FSM, you could rerun the simulation and the results would likely be different. Even with a biased random walk, the results on each iteration may be a little more predictable but still not the same. If the result were always the same, the system would be deterministic and no fun to analyze. It's the differences between simulation results that we're interested in analyzing. Repeated stochastic simulation is the defining characteristic of a Monte Carlo simulation, so in the next section, we'll complete our algorithm by defining how we want to run each test and collect the results in a meaningful way. Once we have the final piece to the puzzle, we'll start using our Monte Carlo simulation to predict some possible future states for our social network.

Monte Carlo Simulation

We can illustrate the relationship between random walks and Monte Carlo simulations with the simple example of flipping a coin. If we flip a coin and it lands on heads, what have we learned about the coin? Well, we've learned that with an extremely small sample size of 1, the coin lands on heads. Now, how useful do you think this information is for making predictions about the result of future coin flips? Could you predict if this is a fair coin or a trick coin? The answer is no, you couldn't. This single result is not very useful—not yet, anyway. To get a clearer picture, we'd need to repeat this test a reasonably large number of times and record each outcome. Suppose we flip the coin 99 more times and it always lands on heads. This is way outside the expected result of roughly 50 percent, so we could state the coin is definitely not fair.

This situation is similar to the relationship between random walks and Monte Carlo simulations. Monte Carlo simulations are a subset of *repeated-sampling* algorithms, which repeat a test some large number of times to gather statistical distributions. What makes a Monte Carlo simulation different from other repeated sampling algorithms is that it uses repeated random walks to simplify simulating complex interactions within a state machine over time. The random walk through the FSM acts like a single test—a coin flip, a space walk, or some other singular occurrence. The Monte Carlo algorithm then adds a layer to repeat this test over and over to collect the large sample size needed to make accurate predictions about future outcomes.

One predominant use for Monte Carlo simulations is in the field of General Game Playing (GGP). The goal for GGP researchers is to find a generalized algorithm that can play any arbitrary but well-defined game. Think about a system like Deep Blue or the more recent Alpha Go, but designed to play chess and *Go*, as well as backgammon, tic-tac-toe, *Risk*, *Battleship*, and so on. This realm of study extends to single-player games (so-called puzzle games) like *Tower of Hanoi* as well. The automated system, called the *player*, needs to decide on the next valid move from a list of potential moves. This process is known as *goal-oriented planning*. In a state machine with a large number of potential states (chess matches, for example), it's prohibitive to exhaustively search the options to conclusion. Instead, a player needs a strategy to quickly weigh possible options to identify advantageous ones. Monte Carlo simulations are one option that researchers have used with some success[4] by reducing each game to a limited-length random walk through potential game states, then repeatedly testing the outcome of these walks for some goal condition.

As I've mentioned, security often comes down to one researcher's offensive knowledge against another's defensive knowledge. Game theory would label this a zero-sum multiplayer scenario. The term *zero-sum* refers to the case that, for one player to win points, the other player must lose an equal amount of points. Simply put: if you win, I must lose, and vice versa. Chess is the most famous example of zero-sum games, but we also see these conditions in a lot of adversarial interactions like security. For me to bypass your security, your security must get bypassed. For your security controls to block me, my attacks must fail. There are already schools like Stanford University that teach game theory as a way for humans to analyze their security posture. There are also researchers using game theory to model attack and defense scenarios.[5] It seems to me that programmatically applying game theory within information security research is a natural progression from the tools available today, and the simplest place to start that process is with Monte Carlo simulations.

Of course, this simplicity can come at a cost. Monte Carlo simulations can miss obvious advantageous decisions due to the random nature of selection. You can tune the accuracy of the model a bit by adjusting the number of random walks, as well as the maximum length of each random walk. Figure 6-3 shows an example Monte Carlo simulation of random walks like the one in Figure 6-2.

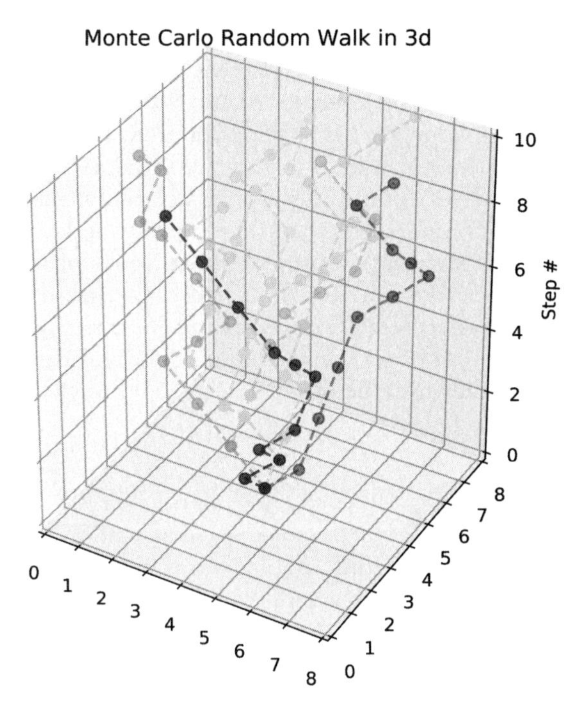

Figure 6-3: A random walk Monte Carlo simulation

Each walk is shaded differently so you can tell where they overlap. They all start from the same location but then take unpredictable paths.

NOTE *The code to generate the random walks and 3D plot in Figure 6-3 is in the 4th cell of the* State_Machine_Graphs.ipynb *notebook.*

The Monte Carlo simulation we'll look at is an algorithm that relies on *k* random walks of length *n*, through a state machine *M*, to obtain the result list ζR. The result is a list of terminal states from each random walk performed:

$$\zeta R = [S^0{}_{(Tn)} \ldots S^k{}_{(Tn)}]$$

For convenience, I also output the path traversed by each random walk:

$$\zeta T = [T^{(0)}_{(n)} \ldots T^{(k)}_{(n)}]$$

Choosing values for *n* and *k* is an equal mix of domain knowledge, statistical theory, and art. For *n*, we need to choose a value large enough to allow our model to reach interesting outcome states without creating a bunch of repetitive data. For state machines with a large number of potential transitions and states, you may need to pick a value for *n* that balances long enough paths with a reasonable program runtime. Our state machine has a small number of potential transitions and will tend to reach a terminal state fairly quickly, so a small value between 10 and 20 steps will suffice.

Choosing a good value for k is largely related to the number of potential states in the machine. You want to run the simulation as many times as it takes to collect statistical data to support a claim about the outcome, so the more possible outcomes there are, the more times you'll want to run the simulation. When you move a project like this into production, you can use statistical methods to calculate the exact sample size required to justify an empirical claim, called *sample size determination*. Here, our simulation has a relatively small number of terminal states and we're only trying to prove out the system, so somewhere between 10 and 25 runs will suffice for testing purposes.

Simulating Social Networks

To answer our research questions about a social network, we're going to write our own *Matrix*-like world, where simulated users live out their digital lives according to the rules of the system we put in place. The rules we choose represent all the decisions a user can make in our simulated world. I base the rules I use on the observations already present in the data (for example, which users have communicated in the past, and on what topics), as well as a simplified version of some link prediction theory published in 2009.[6] *Link prediction theory* attempts to describe how edges in a graph have formed previously and use that information to predict how they'll form in the future.

Our goal in designing the rules for the FSM is to accurately simulate which users might form connections, dissolve their association with other users, or pass information along to their connections. We'll then look at how we can enhance the simulation by adding an adversary who is working to disrupt the network. This allows us to move into the realm of "what if" simulations. What if the head of HR suddenly leaves the company? What if the router in the office crashes? You'll start seeing chances to apply simulations everywhere. After reading through this implementation, think about the rules and assumptions we've built up and how you might improve the simulation with more realistic constraints and behaviors.

Modeling User Interaction

To answer the question "How far is information likely to spread from a given node?" we can simulate a message q propagating through the network by being transmitted from one user to another, and determine how many users are likely to receive the message. We'll model the user interactions by generating biased random walks for the message to move from node to node. Assume, for the moment, that only one copy of the message can exist at a time. (See "Modeling Information Flow" for handling multiple copies of a message.) Think of this like a budget report making its way around an office. As each employee reads the report, they decide whether to forward it on to one of their coworkers. Because the report is sensitive, no one is allowed to make copies, so only one person can be holding the information at any given time. By selecting a starting node and allowing the message to propagate probabilistically, we can simulate possible paths the report is likely to follow, then count the unique nodes that eventually received the message. The average count of unique nodes after all the walks have been completed can be seen as the number of nodes likely to receive the information originating from the selected starting node.

For a Monte Carlo simulation, you must define the state machine that will act as the core of the system. The social network graph nodes (the users) represent states the message can occupy in the FSM. Edges indicate potential transitions between states (which is based on past communications between users). In the beginning of our simulations these will remain static, and we'll be examining the network as it exists at the current point in time. (In the proof of concept for this chapter, the edges will change to simulate users making and severing connections within the network.) Finally, the input alphabet, which defines valid actions for the available transitions, models interactions between nodes (for example, one user passing the message to another). Defining inputs and transitions for the FSM is similar to defining what are valid choices and when. For the first question, which deals with information transmitted between nodes, the input alphabet is [*Send*, *Pass*], representing the two actions a user may take when they receive a piece of information.

To start, we'll define the initial state S_0 as the node with the highest outdegree holding the message, so the simulation has the best chance of reaching a large number of unique nodes on different simulations. Later, we'll measure the effect of starting from different nodes. The node holding the message at any given moment is $u(q)$.

At each step in a random walk, the node $u(q)$ uniformly selects one of two possible inputs Ξ = [*Send*, *Pass*]. The a priori probability of an input being selected is:

$$Pr(\Xi_{(i)}) = \frac{1}{|\Xi|} = \frac{1}{2} = 0.5$$

$Pr\ (\Xi_{(Send)})$ denotes the probability of *Send*. If $u(q)$ chooses *Send*, q is passed to a uniformly selected neighbor of $u(q)$ (I still denote the neighbors of a node as $\Gamma_{(u)}$). If *Pass* is chosen, $u(q)$ does nothing for that step.

The a priori probability of a given neighbor v being selected is:

$$Pr(v^{(q)}) = \frac{1}{|\Gamma_{(u^{(q)})}|}$$

Here $v \in \Gamma_{(u^{(q)})}$. Simply put, this means the starting probability for each neighbor is equal. The larger the number of neighbors a node has, the lower the probability of any one neighbor receiving the message next. For example, if $u(q)$ has three neighbors ($|\Gamma_{(u^{(q)})}|=3$), then $Pr(v^{(q)})=1/3=0.33$. You could also write this as a conditional probability ($Pr(A|B)$):

$$Pr(v^{(q)} \mid Send) = \frac{Pr(\Xi_{(Send)}) \wedge Pr(v^{(q)})}{Pr(\Xi_{(Send)})} = \frac{0.5 \times 0.33}{0.5} = 0.33$$

The formula states that the probability of a particular neighbor receiving the message, given the input is *Send*, is 0.33, or 33 percent.

The overall probability of a message propagating without presuming that *Send* is the selected input is defined in the numerator of the previous equation. Intuitively, you can think of this as the probability of each event occurring in isolation. More properly, the independent probability of a message propagating forward to a given neighbor of $u(q)$ (assuming three neighbors) is:

$$Pr(u^{(q)} \rightarrow v) = Pr(\Xi_{(Send)} \wedge v^{(q)}) = 0.5 \times 0.33 = 0.165$$

Before we generate random walks, we need to set up the simulation, as shown in Listing 6-1.

```
❶ XI = ["send", None]
❷ k = 10
  n = 10
  out_deg = G.out_degree()
  valkey_sorted = sorted(out_deg, key=lambda x: (x[1], x[0]))
❸ S0 = valkey_sorted[-1][0]
```

Listing 6-1: The initialization code for the Monte Carlo simulation

The code relies on the graph *G* being populated using the method back in Listings 5-2 and 5-4. Assuming the graph *G* has already been populated, we start with the input alphabet `XI`, which represents *Send* as expected and *Pass* using `None` ❶; the number of simulations, `k` ❷; and the number of steps in each simulation, `n`. To set the start state `S0` ❸, we select the node with the highest out-degree.

Continuing from Listing 6-1, Listing 6-2 shows a deterministic, uniformly random implementation of the message-passing Monte Carlo simulation algorithm.

```
  from random import choice
❶ R = []
❷ for i in range(k):
    ❸ message_at = S0
      Tn = []
    ❹ for j in range(n):
          if choice(XI) is not None:
            ❺ gamma_uq = list(nx.neighbors(G, message_at))
            ❻ if len(gamma_uq) > 1:
                  vq = choice(gamma_uq)
                  Tn.append((message_at, vq))
                  message_at = vq
              elif len(gamma_uq) == 1:
```

```
                vq = gamma_uq[0]
                Tn.append((message_at, vq))
                message_at = vq
        ❼ else:
                conc = "Message terminated at node %s in %d steps"
                print( conc % (message_at, len(Tn)))
                break
    ❽ R.append((message_at, Tn))
  tot = 0
❾ for end, path in R:
      uniq = unique(path)
      tot = len(uniq) - 1
❿ print(S0, (tot / len(R)) / (len(G.nodes.keys()) - 1))
```

Listing 6-2: A deterministic message-passing Monte Carlo simulation

NOTE *This code is in the 3rd cell of the* MonteCarloSimulation.ipynb *notebook in the chapter's supplemental material, wrapped in a function called* `run_sim_1`.

First, we initialize the results list `R` ❶, and then we use nested `for` loops ❷❹ to perform `k` random walks with up to `n` steps in each.

Each walk begins with the message at the node `S0` ❸. At each step, we gather the neighbors for the currently selected node ❺. If exactly one neighbor exists, this neighbor is automatically selected. However, if more than one neighbor exists ❻, we select one uniformly at random using the `choice` function, then update the `message_at` variable. If the message ever reaches a node with no out-degree ❼, we record the node as the result and conclude the walk with `break`. At the end of each walk, we append the terminating node `Tn` to the results list ❽.

We summarize the likely *information flow distance*, or *IFD* ($IFD = \frac{|T_{(n)}| - 1}{|V| - 1}$), as the mean number of unique nodes (disregarding the starting node) in each path ❾, then normalize the IFD by the total number of nodes in `G` (again, disregarding the starting node) ❿ and print it out to the screen. The `unique` function simply takes the path and reduces it to only unique entries. (You can see how I implemented it in the 2nd cell of the *MonteCarloSimulations.ipynb* notebook in the chapter's supplemental materials. You can also choose to use one of the library's versions of the code, such as NumPy's `unique`.)

NOTE *The absolute value of a list with repeated items is the length of its set equivalent; that is, $|T_{(n)}|$ counts only the unique elements in the random walk, even if the message passes to the same user more than once.*

If you run the code in Listing 6-2 a few times, you'll notice the output isn't consistent. While S_0 is deterministic, the route from there is stochastic. You could edit this model to be entirely deterministic by replacing the `vq = choice(gamma_uq)` call with a deterministic selection method, such as always passing q to the neighbor of $u(q)$ with the highest out-degree. This would be a good option to model a specific behavior pattern that's known in advance.

To implement the simulation with a biased walk instead of a uniformly random walk, you can add another element to `XI` that is the same as one already present. By doing so, you change the relative probabilities of each input variable being selected. For example, adding another `"send"` to the list will weight *Send* to twice as likely as *Pass*:

$$Pr(\Xi_{(Send)}) = \frac{2}{3} = 0.66$$

For more fine-tuned control over the probability (bias) of each input, switch out the simple `choice` function for one that can process a dictionary of `{action: probability}` definitions. (See the proof of concept at the end of the chapter for an example.)

Congratulations, you've now defined your first predictive model using Monte Carlo simulations! This is a simplistic model where we rely on uniform selection for randomness and some basic actions, like send and pass, to describe what might occur to some arbitrary message on our network. Try starting the message from different users and see how it impacts the number of steps the message travels and where it ends up. We refer to this as a *naive model*, because we didn't include any specific information about the history of the network, message contents, or user preferences. We assume each node is equally likely to send any message to any other node it can contact. While simplifying assumptions like this make the code easier to write and interpret, they do so at the expense of accuracy. In the next section, we'll extend our model to incorporate more details about the message and users to more accurately predict the probable flow of information given what we've already witnessed about data flow in the network.

Modeling Topic-Based Influence

To answer the question "Which nodes are being influenced by other nodes?" we'll extend the investigation of topic-based influence from Chapter 5. Recall that we previously weighed each user's potential interest in a topic by measuring their interaction with other messages containing the same topic, using the Hyperlink-Induced Topic Search algorithm (HITS). If we reframe our current model with respect to a given message topic, like the environment, we can incorporate hub and authority information into our state machine model to control the information exchange probability. In this case, we'll use a user's HITS score to determine the probability of a message being reblogged based on the message's content, instead of just blindly assuming all messages have the same probability for all users all the time.

Modeling message propagation in this fashion assumes that a user is more likely to reblog a message similar to a message they have reblogged previously. Users who have reblogged posts involving a given topic get a higher authority score for that topic than those who haven't, which translates to a higher probability of receiving a message about that topic. If you think about the content you see on people's social network feeds, you'll probably see a fairly common theme among the information they share and reshare (this is one of those assumptions you may want to challenge later). Some people choose to share business news; others, arts and entertainment; and still others, security.

Let's update the previous implementation to compare the spread of different message types (*qx*) so we can examine the interests of different users and predict what messages they're most likely to reblog in the future. If you were designing a viral message attack for this network, it would make sense for you to examine different topics and choose the one with the highest probability of propagating farthest through the network. From a defensive perspective, you can flip this analysis and track a malicious message back to the probable source. We'll be keeping the same definition of influence between users (so a user reblogging a message is influenced by that message to some degree), but instead of using the *Send* action, our nodes will be selecting messages to reblog. Nothing about the action changes, so I've opted to keep the name, but renaming it might help to keep the direction of influence clear in your model.

Listing 6-3 shows the code to run a topic-based message-passing Monte Carlo simulation:

```
import graph_funcs as ext
qx = "environment"
❶ hG = term_subgraph(qx, post_df)
hub_scores, auth_scores = nx.hits(hG, max_iter=1000, tol=0.01)
hub_max = max(hub_scores.values())
S0_i = list(hub_scores.values()).index(hub_max)
❷ S0 = list(hub_scores.keys())[S0_i]

for i in range(k):
    uq = S0
    Tn = []
    for j in range(n):
      ❸ send_msg = ext.hub_send(hub_scores[uq])
        if send_msg:
          ❹ vq = ext.scored_neighbor_select(hG, uq, auth_scores)
            if vq is None:
                conc = "Message terminated at node %s in %d steps"
                print( conc % (uq, len(Tn)))
                break
            else:
                Tn.append((uq, vq))
              ❺ uq = vq
    R.append((uq, Tn))
ended_at = {}
❻ for end, path in R:
    if end in ended_at.keys():
        ended_at[end] += 1
    else:
        ended_at[end] = 1
return (S0, ended_at)
```

Listing 6-3: A topic-based message-passing Monte Carlo simulation

This code prints a tuple of (`S0, {node: termination_count}`), using the same values for `k` and `n` defined in Listing 6-1, as well as the `term_subgraph` function ❶, which is based on code you'll see later in Listing 6-5.

NOTE *This code is in the 4th cell of the* MonteCarloSimulations.ipynb *file in the chapter's supplemental materials in a function named called* `run_sim_2`.

In this simulation, S_0 ❷ is the node with the highest hub score for the selected topic ($S_0 = max(hub(qx)(G))$), and $Pr\ (\Xi_{(Send)})$ is the hub score of $u(q)$ for the message type x: $Pr\ (\Xi_{(Send)}) = hub_{(qx)}(u)$. The `hub_send` function (defined in the *graph_funcs.py* file, provided in the book's supplementary materials) takes the hub score of `uq` and returns whether `uq` passes the message on ❸. The `hub_send` function is based on another function, `weighted_choice`, which is also included in the *graph_funcs.py* file. There are still only two possible actions in Ξ, so the probability of *Pass* is equal to 1 minus the probability of *Send*: $Pr\ (\Xi_{(Pass)}) = 1 - Pr\ (\Xi_{(Send)})$. The probability for selecting a given neighbor is the normalized authority score for that neighbor, given the message type qx ($Pr(v(qx)) = auth(qx)(v)$).

If the message is sent, we select the neighbor using the `scored_neighbor_select` function (also defined in the *graph_funcs.py* file and based on the `weighted_choice` function) ❹. If a neighbor is returned, we add the edge between the sender and recipient to the path `Tn` and update the message location ❺; otherwise, we terminate the simulation with the `break` statement.

If we assume a message ending at a node imparts some influence, we can count how many times a message ends at a particular user and claim that the node with the highest count is most likely to be influenced by the given message type from the given user. This intuitively means that the user would likely end up reblogging the message at some point. All roads lead home, so to speak. To find this node, we loop over the ending locations and tally the results to build the `{node: termination_count}` dictionary ❻, then print the results. This constitutes one run of the Monte Carlo simulation.

We want to collect several runs and average the results for the most accurate predictions, so we wrap the code in a function definition called `run_sim_2`, which will take in the topic list and the Mastodon post data as parameters. (You can see the `run_sim_2` function in the 4th cell of the *MonteCarloSimulation.ipynb* notebook.) Finally, we return the source node and the dictionary containing the users at whom the message terminated, so we can collect the results until we're ready to analyze them. Let's call this newly defined function in a loop to collect a reasonable sample size. Listing 6-4 shows how to collect the samples and average them for the final output.

```
  all_runs = {}
  started_at = ""
❶ for run_i in range(0, 10):
    ❷ started_at, results = run_sim(["environment"], post_df)
    ❸ for ks in results:
          if ks in all_runs.keys():
              all_runs[ks] += results[ks]
```

```
        else:
            all_runs[ks] = results[ks]
❹ for node in all_runs.keys():
    if node != started_at:
        print("%s influenced %s an average of %.2f times" % (
            started_at, node, all_runs[node]/10
        ))
```

Listing 6-4: Averaging the Monte Carlo simulation results

Here are the results from averaging 10 runs of the simulation:

```
gutierrezjamie influenced iwatkins an average of 1.80 times
gutierrezjamie influenced hartmanmatthew an average of 2.20 times
gutierrezjamie influenced shannon42 an average of 0.90 times
gutierrezjamie influenced daniel99 an average of 0.70 times
gutierrezjamie influenced garciajames an average of 1.00 times
gutierrezjamie influenced grosslinda an average of 0.30 times
```

Using these results, we could claim that the user `gutierrezjamie` is most likely to influence `hartmanmatthew` on the topic of environment. What's important here isn't the numbers themselves but the relative sizes of the numbers, so you might also conclude that `shannon42` is three times more likely than `grosslinda` to end up reblogging the message. Of course, this is just the result of one small group of simulations. Ten simulations on a topic as complex as information flow and influence is hardly definitive. To strengthen this claim of influence, repeat the simulations some large number of times by increasing k (using the statistical method mentioned earlier) and average those results. In general, the more possible outcomes for the simulation, the more times it should be run. There's a point of diminishing returns to this, though. You'll want to experiment with different simulation counts and lengths by updating the values for k and n, respectively.

The last part of the code we're going to examine is the `term_subgraph` function, which we called in Listing 6-3. The function in Listing 6-5 takes in a term of interest and searches the underlying data to find all relevant posts.

```
def term_subgraph(term, df):
    dat_rows = df[df["text"].str.contains(term)]
    dat_replies = df[df["in_reply_to_id"].isin(dat_rows["id"].values)]
    hG = nx.DiGraph()
    for idx in dat_replies.index:
        row = dat_replies.loc[idx]
        hG.add_edge(row["in_reply_to_screen_name"], row["user_screen_name"])
    return hG
```

Listing 6-5: Defining the subgraph based on terms

The function takes in the term we're interested in searching for and the `post_df` `DataFrame` object we defined previously. Using the `str.contains` function, we filter the data down to only rows whose text column contains the search term. We then collect the replies to these posts by searching the `in_reply_to_id` column for any relevant post IDs, storing them in a `DataFrame` called `dat_replies`. Next, we define the `DiGraph` object that will hold the resulting graph data and store it in a variable named `hG`. We loop over the `dat_replies` index list, and, for each entry, we look up the row associated with the index. We use the row's `in_reply_to_screen_name` and `user_screen_name` to create an edge in the graph, showing the direction of influence on the topic of interest. Once we've completed the loop, we return the completed subgraph.

Now that we've defined all the basic code we'll need, we can start to improve upon our simple model. In the next section, we'll cover how to make our message behave more realistically through resource allocation.

Modeling Information Flow

So far, our simulation treats the message like a single object moving from node to node, like a package being delivered to an address. But what about the cases where a message may be transferred to multiple users simultaneously? Our model is fine for single copies of messages, but it'd be nice if we could find a way to model the information more intuitively as flowing through the network instead. Think about it like this: you don't send one birthday party invitation and ask each invitee to pass the message on to the next person on the list; you send multiple invitations to the people you want to attend. Each invitee may then invite another person to go to the party with them, so the message spreads even further along the network simultaneously. To model this type of information flow, we need to improve our state machine to treat the message as if multiple copies exist.

To simulate the case where more than one copy of q can exist, we can reformulate message passing as a question of resource flow within the network. By doing so, we can figure out how much information two people have communicated in the past and use that as an indicator of how much they may communicate in the future. *Resource allocation (RA)* is a model that was first posed to describe the nonlinear correlation between airport connectivity and travel capacity.[7] We'll be using the same principle as a way to quantify the quality of information exchange as the message spreads through the network.

Generally speaking, RA describes the potential flow of resources between two nodes (u, v) where v is not a neighbor of u, but they're connected by a directed path $(v \notin \Gamma_{(u)} \wedge \rho(u \rightarrow v) \in E)$. Supposing a node u in a directed graph has one unit of resources to distribute evenly among all its direct neighbors, the resources allocated to any member of the network is the sum of the resources at the end of each path between u and v:

$$RA_{(u \rightarrow v)} = \sum_{w \in \Gamma_{(u \rightarrow)} \cap \Gamma_{(v \leftarrow)}} \frac{1}{|\Gamma_{(w)}|}$$

You can think of this value as the importance of node v in the case of distribution for node u. If $|\rho(u, v)| > 2$, this process is repeated for all nodes between, until some amount of resource reaches v. Therefore, you might instead wish to think of this value as the amount of resources u provides to v through the distribution network.

As a concrete example, suppose you're investigating a criminal organization that sells counterfeit goods it purchases from a forger. The boss of this hypothetical organization buys 100 boxes of knock-off handbags (the initial amount of resources at S_0). He then distributes the merchandise to his top 4 lieutenants by dividing the 100 boxes into 25 boxes for each. Finally, each lieutenant divides their 25 boxes among their street corner shops. If each lieutenant has connections to 5 storefronts, each store would get 5 boxes. If the crime boss were to lose one of these stores, the loss would account for only 5 percent of his inventory. This is a very simplistic model that assumes each node and path can evenly carry the resource in question. However, that's not always the case.

Formally, the $\frac{1}{|\Gamma(u)|}$ portion of the previous formula is known as the *flow function*, which models a specific type of behavior for passing or receiving resources. Using this flow function, the resource gets divided evenly among all the neighbors of the node u, the same as the boxes of counterfeit goods. There are a few different flow functions built into NetworkX. Unfortunately, they're not implemented for directed graphs as defined here. As you shift from research to applications, you'll often be responsible for extending your code libraries with missing definitions like this. The *graph_funcs.py* file includes the code for directed resource allocation, so you can use it to experiment.

By combining the historical analysis of the HITS algorithm with the simultaneous flow of resource allocation, we can create a respectable model, capable of simulating user behavior based on previous observations. You should be able to build on this framework of state machines and Monte Carlo simulation to model all sorts of interesting phenomena, not just in social networks but throughout the topic of information security as a whole.

In the next section, we'll move on to a proof-of-concept application that will take us deeper into applied game theory and Monte Carlo simulation by simulating an adversarial face-off on our social network platform. Let's get ready to rumble!

The Proof of Concept: Disrupting the Flow of Information

The final question for this chapter—"What links could be severed to disrupt the flow of information between two nodes?"—is a very interesting security topic. There are many scenarios in which disrupting the flow of information to a particular subset of nodes could be catastrophic. Imagine a hospital tied to a single source of electricity. To disconnect any outlet in the hospital from electricity, you'd only need to sever the single link between the hospital and its power source. This is a *single point of failure*, and to avoid it hospitals deploy multipoint connections to the power grid and install backup generators for more severe disruptions. Many home networks suffer from this design flaw as well. To sever all the connected devices behind the router, you simply need to sever the connection forward of the router. In social networks, like businesses, failure points like these occur regularly. Companies often have people known as "linchpin employees"[8] who fill roles no other employees can, or possess arcane knowledge the company needs to operate. Linchpin employees inspired the proof of concept for this chapter: Monte Carlo simulations to model the potential to disrupt information flow within an evolving social network.

For the rest of this chapter, we'll be building a simulation where our social network is under attack from a nefarious outsider. We'll use some of the same analysis techniques that gave birth to the modern internet to see how difficult it would be to disrupt our social network. Sometimes it's fun to be the bad guy!

Modeling an Evolving Network

In a static network, you could find the set of edges that, when removed, would separate two nodes (you'll see a method for producing this list in a moment), but that doesn't account for network adaptations, like cross-training another employee in the linchpin employee's arcane knowledge to alleviate a single point of failure. To model an evolving network, we'll mimic a two-player, turn-based game scenario, wherein one player tries to get a message through from a starting user to an end user as their adversary tries to stop the message from reaching the end. To make the game more complex, the network itself evolves on each turn as users reblog messages from other users or disconnect from people who were previous connections. Player 1 acts on the part of the network and all its users, while player 2 acts as the adversarial force. Player 1's goal is to send a message u from a source node (uA) to a sink node (v_Ω). Player 2 seeks to keep this message from reaching the sink node by selectively removing paths from the network. The game play is broken up into three phases: network adaptations, message movement, and finally adversarial movement (in that order).

The game is over if q reaches v_Ω, or when no path exists to complete the transmission:

$$\zeta R = \begin{bmatrix} u^{(q)} = v_\Omega: & 1 \\ \rho\ (u^{(q)} \rightarrow v_\Omega) \notin E: & -1 \end{bmatrix}$$

This equation can be translated into the convenient helper function in Listing 6-6.

```
def check_win(G, uq, omega):
  ❶ if uq != omega and nx.has_path(G, uq, omega):
        return None
  ❷ elif uq == omega:
        return 1
  ❸ elif not nx.has_path(G, uq, omega):
        return -1
```

Listing 6-6: Checking for terminal conditions

This function takes in the graph object, the ID of the node currently holding the message, and the ID of the goal node. If these two nodes aren't the same and there's a path between them ❶, the game isn't over, so the code returns None (as in no winner). If the two nodes aren't equal and there's no path between the current node and the goal node ❸, player 2 has succeeded in isolating the message, and the function returns -1. If the two IDs match ❷, the message has reached the goal node, so the function returns 1.

Moving the Message Through the Network

The second helper function, weighted_choice, shown in Listing 6-7, will be used for weighted random selection of the next node to a receive the message.

```
def weighted_choice(scores):
    totals = []
    running_total = 0
  ❶ for w in scores.values():
        running_total += w
        totals.append(running_total)
  ❷ rnd = random() * running_total
    for i in range(len(totals)):
      ❸ if rnd <= totals[i]:
            key = list(scores.keys())[i]
            return key
```

Listing 6-7: Weighted random selection function for biased walks

The input parameter scores is a dictionary of {item: weight}, giving each item that may be selected and its weight. (The weights do not need to sum to 1; only the relative size of the values matters.) The totals list partitions the real number space between 0 and the sum of the weights ($0 \leq rnd \leq \sum_{i=0}^{j} w_{(j)}$) into bins proportional in size to the weight of the item they represent, by adding each item to a running_total, then recording the running total after each item is added ❶. The sum of all the weights then scales a random value ❷ to fall into one of the bins, and the bin determines which item is selected ❸. Items with larger weights map to larger bins, meaning the items are more likely to be selected, hence "weighted random selection."

As a concrete example, take an input dictionary {"A":1,"B":2,"C":3}. After the first loop executes, the totals list contains [1,3,6] and the running_total is 6. The random real value rnd (between 0 and 1) is selected using the random function, then multiplied by running_total ❷ to produce the percentage of the weight randomly selected. The random value 1.0 means the maximum weight, in this case:

$$1.0 \times \sum_{i=0}^{j} w_{(j)} = 1.0 \times 6 = 6$$

We can verify that the break points accurately reflect our input weights by calculating the amount of space on the number line assigned to the key, called its *key space*. These should equate to 1 / 6 = 0.166, 2 / 6 = 0.333, and 3 / 6 = 0.5 for keys A, B, and C, respectively. We find the key space by subtracting the key's lower selection boundary from its upper selection boundary. To select key `A`, `rnd` must be lower than or equal to approximately 0.166 (0.166 × 6 = 0.996). To select key `B`, `rnd` needs to be between 0.166 and 0.5 (0.5 × 6 = 3), which means the key space for `B` is (0.5 – 0.166 = 0.333). We can divide `B`'s key space by `A`'s to get a relative size comparison (0.333 / 0.166 = 2.006), which means the key space for `B` is twice the size of the one for `A`, just as we requested. Finally, `rnd` needs to be greater than 0.5 and less than or equal to 1.0 for key `C` to be selected. The key space for `C` is (1 – 0.5 = 0.5). You can continue the key space logic to prove that the space provided to `C` is three times the space provided for `A` (0.5 / 0.166 = 3.0) and one and a half times larger than the key space for `B` (0.5 / 0.333 = 1.5). I hope this helps to illustrate how the values in our input dictionary control the size of the key space created during the random selection process. We'll be relying heavily on the `weighted_choice` function during our proof, so it's worth taking the time to understand it in detail.

Measuring the Amount of Information Flow

Some connections between nodes in a network may be capable of carrying more information than others. In a social network, for example, some members may be more effective at spreading information, like a linchpin employee at a company. The amount of information flow between two nodes in a social network isn't straightforward to measure, and depends on your research question. For the purposes of our simulation game, the edge capacity is the number of characters in a given post (maximum 500, at the time of writing).

NOTE *The number of characters in the text body associated with each edge just measures raw data flow between nodes, and naively assumes all data translates equally to information. See the chapter's summary for recommended reading on other heuristic measures for measuring information in text.*

By adding an attribute named *capacity*, which represents the maximum amount of information that can be transmitted along the particular edge in one unit of time, to the edges in *E*, we can compare the effect of removing different subsets of edges on the overall flow and capacitance of the network using the *max-flow, min-cut theorem.*[9] We'll take a deeper look at this theorem when we improve player 2, but for the moment just know that it allows us to model a resource that gets spread out across a network, rather than moving from point to point as we've seen previously.

Now is a good time to step back and remember why this matters to us. Our ultimate goal is to test how difficult it would be for an adversary to significantly disrupt the communication of the network. The max-flow, min-cut theorem gives us the information needed to test if two nodes can still communicate (because there's still a path between the two nodes). It also helps us determine what cuts are more or less advantageous for the adversary, as well as giving them a way to quickly judge their options. An attacker who knows the max-flow, min-cut theory will likely have a much higher chance at sabotaging the network than one who attacks random communication channels.

We'll examine this hypothesis by implementing two versions of the adversary in the following game and comparing the damage they can achieve.

How the Game Works

The game is actually very simple. The objective for player 1, the white hat, is to get a message from a node on one side of the network to a node on the other side. To achieve this they have the entire network at their disposal. On each turn, they will move the message around the network trying to reach the sink node. They win if the message successfully traverses the network from the source node to the sink node. On the other side of the virtual table is player 2, the black hat. Their job is stop that message, at any cost. On each turn, they'll select an edge to remove from the network. Player 2 wins if they successfully disconnect the network so that there's no way the message can reach the sink node.

Evolving the Network

Social networks rarely have a static topography: the links and membership are changing even as you try to measure them. The network adaptation phase models the evolving topography by allowing edges to be created or removed probabilistically. This means that new routes may open up suddenly and old paths may disappear on their own. Neither player can fully trust the network to do what they expect it to. I chose to implement this as part of player 1's turn since they're the network administrator in this scenario. On each turn, player 1 chooses an input action from the input alphabet for every node that's not holding the message ($\forall u^{(\neg q)} \in V \, wrs(\Xi)$, where $wrs(\Xi)$ is the weighted random selection function defined previously). Listing 6-8 shows the weighted input alphabet, which includes the creation and dissolution of edges with `connect` and `disconnect`, or the option to `pass` as before.

```
XI = {
    "connect": 2,
    "disconnect": 1,
    "pass": 2
}
```

Listing 6-8: Weighted inputs for nodes without the message

The weights in `XI` describe the tendency of the network over time. These values create a scenario where the network is likely to grow over time since `connect` and `pass` both have an individual weight of 40 percent (2 / 5 = 0.4), a combined 80 percent of the selection space, while `disconnect` has only 20 percent. If `connect` is chosen, the node forms a new edge with (meaning it receives a reply from) another node in the graph, which means we need a way to select the user they connect to. I've chosen to implement this based on *preferential attachment*, the idea that users with many connections are more likely to form new connections than those with fewer connections. In our network, this means that nodes that receive replies from many other users (large out-degrees) are more likely to receive a reply from users who reply to many users (high in-degrees). (Even if the current node doesn't tend to receive many replies, it's still more likely to get a reply from a more active user.)[10]

Formally, the *undirected preferential attachment (UPA)* score for two nodes (u, v) is the product of the length of their neighbors:

$$UPA(u,v) = |\Gamma_{(u)}| \times |\Gamma_{(v)}|$$

To account for the directionality of our network graph, we can define *directed preferential attachment (DPA)* using outgoing neighbors of u and incoming neighbors of v:

$$DPA(u \rightarrow v) = |\Gamma_{(u \rightarrow)}| \times |\Gamma_{(v \leftarrow)}|$$

Listing 6-9 shows the weighted random connection function, which will be called from the player 1 logic shown in Listing 6-12.

```
def wrs_connect(G, u):
    scores = {}
  ❶ for i in range(len(G.nodes.keys())):
        v = list(G.nodes.keys())[i]
        if v == u:
            continue
      ❷ dpa_score = G.out_degree(u) * G.in_degree(v)
        scores[v] = dpa_score
  ❸ return weighted_choice(scores) # Previously defined choice function
```

Listing 6-9: A directed preferential attachment weighted random selection

The function `wrs_connect` takes the graph and the connecting node as input and loops over each ID in the graph ❶ to calculate the DPA score ❷ between the input node and each other node (skipping the input node with `continue`). The `weighted_choice` function uses the `scores` dictionary to pick a node to connect to ❸:

$$v_{conn} = wrs([DPA(u^{(\neg q)}, \neg u)] \forall \neg u \in V)$$

If `disconnect` is chosen, the user disassociates from another user who sends them the least information (as measured by the definition of capacity given previously). The `capacity` attribute of each edge in `G.in_edges` is used in the weighted random selection process to choose a neighbor to disassociate from. Listing 6-10 shows the code to calculate the capacity and selection weight for a given node.

```
def ncap_weights(G, u):
    u_in = list(G.in_edges(u, data=True))
    n_capacity = {}
  ❶ for v,u,d in u_in:
        n_capacity[v] = d["capacity"]
  ❷ Q = 1 + max([itm[1] for itm in list(n_capacity.items())])
  ❸ n_weight = {k: (Q - n_capacity[k]) for k in n_capacity.keys()}
    return (n_capacity, n_weight)
```

Listing 6-10: Capacity weighting for a node

We start by looping over the neighbor data from the set of inbound edges for the node ❶ and collecting these into the n_capacity dictionary. If we hadn't condensed the multiple edges previously (by using a DiGraph instead of a MultiDiGraph), we'd need to sum up the capacity of each in-edge incident to the node first:

$$N^{(u \leftarrow v)}_{capacity} = \{(u \leftarrow v)_{capacity}\} \forall v \in \Gamma_{(u \leftarrow)}$$

We weight edges with lower $N_{capacity}$ entries more heavily in the selection process, then invert the capacities by subtracting them from a modifier $Q = 1 + max(N_{capacity})$ ❷, which results in the weighting formula $N^{(u \leftarrow v)}_{weight} = Q - N^{(u \leftarrow v)}_{capacity}$ ❸.

For example, if $max(N_{capacity}) = 10 \Leftrightarrow Q = 11$, an edge with $N^{(u)}_{capacity} = 10$ will get a weight of 1 ($N^{(u \leftarrow v)}_{weight} = 11 - 10 = 1$), while an edge with $N^{(u \leftarrow v)}_{capacity} = 1$ will get a weight of 10 ($N^{(u \leftarrow v)}_{weight} = 11 - 1 = 10$).

We'll set the return value for a given node to the tuple $(N_{capacity}, N_{weight})$. Both n_capacity and n_weight are dictionaries keyed off of the neighboring node's ID. The wrs_disconnect function in Listing 6-11 uses the n_weight dictionary to select the least informative neighbor to disconnect u from.

```
def wrs_disconnect(G, u):
 ❶ u_in = list(G.in_edges(u))
    if len(u_in) < 1:
        return None
 ❷ caps, scores = ncap_weights(G, u)
    if scores is not None:
     ❸ return ext.weighted_choice(scores)
```

Listing 6-11: A weighted random disconnection function

If there are no inbound edges for the node u (meaning it has no in-degree neighbors to disconnect from) the function returns None ❶, resulting in the same outcome as pass. If more than one edge is found, the capacity scores for all the inbound neighbors are calculated using the function from Listing 6-10 ❷. The key returned from this function is the neighbor to disassociate from ($v_{disconn} = wrs(N_{weight})$) ❸. The edge ($u^{(\neg q)} \leftarrow v_{disconn}$) is then removed from the graph in the player 1 logic in Listing 6-12.

Moving the Message

After the network evolution phase, the game moves into the message movement phase, where the only possible input is send. If an edge exists between the node currently holding the message, $u(q)$, and the goal node, v_Ω, the message passes along that edge and player 1 wins the game. Otherwise, the paths between $u(q)$ and v_Ω are calculated, and the message is passed to the next node along one of these paths, selected uniformly at random.

The code in Listing 6-12 handles player 1's turn, which includes both the network adaptations and message movement phases. The function player_one_turn takes the graph, the node holding the message, and the goal node as parameters and returns the node that receives the message and the new state of the graph.

```
def player_one_turn(G, uq, omega):
  ❶ if G.has_edge(uq, omega):
        return (omega, G)
    caps = [d["capacity"] for u,v,d in G.edges(data=True)]
  ❷ avg_cap = sum(caps) / len(caps)

  ❸ for u in list(G.nodes.keys()):
        if u == uq:
            try:
              ❹ paths = list(nx.all_shortest_paths(G, u, omega))
            except nx.exception.NetworkXNoPath:
                return (uq, G)
          ❺ path = choice(paths)
            pass_to = path[1]
        else:
          ❻ act = ext.weighted_choice(XI)
          ❼ if act == "pass":
                continue
          ❽ elif act == "connect":
                v_conn = wrs_connect(G, u)
                G.add_edge(u, v_conn, capacity=avg_cap)
          ❾ else:
                v_disconn = wrs_disconnect(G, u)
                if v_disconn is None:
                    continue
                G.remove_edge(v_disconn, u)
  ❿ return (pass_to, G)
```

Listing 6-12: The logic defining player 1's turn

If an edge exists between uq and the goal node omega ❶, we pass the message to omega. This will end the turn (and game) with a victory for player 1! Otherwise, we calculate the average capacity of the graph ❷, which will be used as the capacity of any new edges added during network adaptation. This allows the average to change from turn to turn, depending on which edges (if any) were removed during the previous network adaptation phase.

To perform network adaption and message passing, we loop over each node in the graph ❸. For $u(q)$, we attempt to find valid paths between it and the goal node ❹ (the message movement phase). If this attempt fails with an nx.exception.NetworkXNoPath, the function returns and the round ends with a victory for player 2, since the message can't reach the destination. Otherwise, we randomly select a path using the choice function ❺ and pass the message to the first node in this path.

For all other nodes, we select an action using the weighted random function and the XI dictionary defined in Listing 6-8 ❻. If pass is chosen, the code jumps to the next node using the continue keyword ❼. If connect is returned, we use the wrs_connect function shown in Listing 6-9 to form a new edge ❽. Otherwise, disconnect was chosen, so we use the wrs_disconnect function from Listing 6-11 to remove an edge from the graph ❾. Finally, we return the receiving node and the updated graph ❿.

Disrupting the Network

Player 2 then gets to select an edge to remove from the network to try to disrupt the flow of the message. One of the strengths of Monte Carlo simulation is in its ability to compare different strategies over time. To illustrate this, let's compare two strategies for player 2 to achieve their goal. In the first strategy (Listing 6-13), player 2 selects an edge from E uniformly at random:

```
def player_two_random(G):
  ❶ e = choice(list(G.edges()))
  ❷ G.remove_edge(*e)
  ❸ return G
```

Listing 6-13: The player 2 random implementation

This will act as a good baseline, since it most closely resembles a truly random walk. The code randomly selects ❶ and then removes ❷ an edge from the graph; we then return the updated graph ❸.

The results of such a strategy can be seen as a null control, as if an adversary just blindly started removing things with no real concept of what they were impacting. We'll look at the second strategy, where player 2 selects their moves to inflict the most damage using the flow information of the network, after we demonstrate the simulation using this simple strategy and gather our baseline network performance.

Once player 2 has finished selecting the edge to disrupt, the round is complete. If neither player has won, the next round starts and gameplay continues in this fashion until one of the players succeeds in reaching their objective.

In the next section, we'll cover how to select the start and end nodes. In much larger networks (like the ones you're likely to see in the wild), having methods for automating tasks like finding data will save you a lot of manual exploration before running your first simulation.

The Game Objective

Before we tie this all together into a functional simulation, let's look at the `shortest_path_scores` helper function, shown in Listing 6-14, which returns a list of average path lengths for all pairs of nodes that are not directly connected.

```
def shortest_path_scores(G):
    pairs = []
  ❶ for u, v in nx.non_edges(G):
        if u == v:
            continue
        if not nx.has_path(G, u, v):
            continue
```

```
    ❷ uv_paths = list(nx.all_shortest_paths(G, u, v))
    ❸ avg_len = sum([len(p) for p in uv_paths]) / len(uv_paths)
      pairs.append(((u, v), len(uv_paths), avg_len))
  sorted_scores = sorted(
      pairs,
    ❹ key=lambda kv: (kv[2], kv[1], kv[0]),
      reverse=True
  )
  return sorted_scores
```

Listing 6-14: Creating an average-length score for weighted selection

The for loop ❶ calls the NetworkX function nx.non_edges to get a list of all possible node combinations not directly connected by an edge, checks that the two nodes *u* and *v* are different, and checks that one or more paths exist between the nodes. If any condition fails, we skip that pair of nodes with the continue keyword. Otherwise, we use the nx.all_shortest_paths function to make a list of all potential paths between the source and sink nodes at the start of the game ❷, then calculate the average path length ❸ and append it to the pairs list. Once all the pairs have been processed, we sort the results in descending order, based on the average path length first, then the ID of the node *u*, and finally the ID of the node *v* ❹.

In Listing 6-15 we'll combine these scores with the weighted_choice function to randomly select the source and sink pair while favoring pairs that have more, or longer, paths than those with fewer, shorter ones. I chose this method so the simulation has enough routes to make it interesting. You may choose the source and sink nodes based on other parameters in your simulation. You might even extend your simulation to test all possible combinations of source and sink node.

The Game Simulation

Finally, it's time to tie all these functions together into a single cohesive game with the code in Listing 6-15. We'll run the game simulation 25 times, each with a different pair of source and sink nodes. Each run will generate *k* random walks, representing one game between player 1 and player 2 per walk, and tally the number of wins for each player. The average of the *k* scores is the score for the run overall. Using different source and sink nodes, instead of running the same scenario over and over, will give us a better sense of the network as a whole. The code in Listing 6-15 assumes you've already built the graph (using code similar to Listing 6-3).

```
path_scores = shortest_path_scores(G)
k = 10 # Number of random walks
n = 10 # Number of steps in each walk
❶ path_weights = {(p[0][0], p[0][1]): p[2] for p in path_scores}
played = []
for r in range(25): # Run the simulation 25 times
  ❷ selected = weighted_choice(path_weights)
    while selected in played:
        selected = weighted_choice(path_weights) # Avoid repeated selection
```

```
    played.append(selected)     # Track new pair
    alpha = selected[0]         # Source node
    omega = selected[1]         # Sink node
    game_res = []               # Results from k random walks
  ❸ for i in range(k):          # Perform k random walk simulations
        newG = G.copy()         # Copy the graph to maintain the original state
        now_at = alpha
      ❹ for j in range(n):      # Perform at most n steps in each walk
          ❺ w = check_win(newG, now_at, omega)
            if w is not None:
                game_res.append(w)
                break
          ❻ now_at, newG = player_one_turn(newG, now_at, omega)
          ❼ if not check_win(newG, now_at, omega):
              ❽ newG = player_two_random(newG)
  ❾ tally = sum(game_res)
    avg = tally / len(game_res)
    print("\t Average %.4f" % avg)
```

Listing 6-15: The main game simulation function

Before running the group of simulations, we get the list of average path lengths between nodes with `shortest_path_scores` (Listing 6-14) ❶, convert the average path lengths returned into a list of `path_weights` (meaning nodes with longer average shorter path lengths get weighted higher), and select a pair of nodes (which is the key returned from `path_weights`) ❷. If that pair of nodes and associated paths have already been used in a simulation (tracked by the `played` list), we select another. From the selected pair and path, we set the source and sink nodes, `alpha` and `omega`.

Once we've found a valid source and sink pair, we perform the *k* random walks. Each iteration of the `for` loop ❸ constitutes one complete game, played on a copy of the graph (`newG`) to maintain the original topology between matches. Each *n*-step random walk ❹ generates up to *n* turns for both players ❻❽ and checks the win condition at each phase ❺❼. The result of each game is appended to `game_res`. Each iteration of this loop counts as one complete turn cycle within a game.

Once the *k* walks are complete, we tally the wins (1 point for a win by player 1 and –1 point for a win by player 2) ❾ by summing `game_res`, then take the average `tally` as the overall score for the *k* walks.

Run the code in Listing 6-15 to see the result of 25 simulations. The averages produced by each test (the outermost `for` loop) may vary wildly, as you can see from this snippet:

```
Average 0.7600
Average 0.2800
Average 0.6000
Average 0.1200
Average -0.5200
Average 0.6800
Average -0.7600
--snip--
```

```
Average 0.9200
Average 0.8400
Average -0.3600
Average 0.5200
Average 0.6000
Average 0.2000
Average -0.5200
```

A score of `0.0` means both players won the same number of matches. Positive averages indicate player 1 won more often than player 2. The closer to `+1` this gets, the more heavily the matches favored player 1. The opposite is true as the score moves below 0. An average score of `-1` indicates player 2 won every match.

The final step is to summarize the result from all the tests. We can do so by summing the individual averages and then dividing by the number of tests. We'll call this the *population mean.* The benefit of the population mean is twofold. First, it summarizes all the tests into a single number you can interpret, rather than a list of test results. Second, the population mean should be relatively stable compared to the values observed between individual run results. If we rerun the code, we'll get different individual test results. The population mean should be relatively stable, though.

When analyzing the model three times, I got the population averages 0.2160, 0.2320, and 0.1808. Of course in statistics we deal with uncertainty, so a better measure of the population mean is the numeric range we believe the actual population mean will fall between, given some desired level of confidence. To do this, we use the `scipy.stats.t.interval` function and pass in the results from our simulations and our desired confidence interval (called the *alpha parameter*) as a float. The result is a tuple containing the lower and upper bounds within which we can predict the true population mean will fall. For example, I ran the simulation 6,250 times and I can say with 95 percent confidence the true population mean for the simulation (as it's currently configured) will be between 0.1078 and 0.2225, which means there is a slight advantage to player 1. The current design seems to slightly favor the defender because of its growth and the lack of intelligence from the adversary.

Now that we've established a baseline performance for our network, let's see if we can improve the adversary's chances by letting them observe the network and pick which routes to sever. We can then compare the results of the two simulations (in terms of predicted population means) and see if our changes significantly impact player 2's chance at disrupting the network.

Improvements to Player 2

Let's see what happens if we give player 2 a little more intelligence. In this version, player 2 uses the updated graph and current message position to remove an edge that's important to the path between the message position and the goal (a relatively intuitive strategy for human players).

To codify this strategy, player 2 will use the max-flow, min-cut theorem. A max-flow, min-cut analysis was one of the driving forces behind the creation of the modern TCP/IP internet. The protocol breaks messages into little chunks called *packets* and then chooses different routes for different parts of the transmission based on response times and carrying capacity. The basic idea of the design is that someone would have to take out a large percentage of the network before they could disconnect two distant nodes from each other. The list of nodes that would need to be removed is known as the *cutset*. The adversary in our simulation game will take advantage of the cutset information to carry out the exact type of attack Paul Baran, the inventor of packet-switching networks, was concerned with in his research (*https://www.rand.org/about/history/baran.list.html*)—that is, the selective targeting and removal of communication channels to disrupt the network.

In short, the max-flow, min-cut theorem tells us two key pieces of information. First, the max-flow portion describes the maximum amount of resources that can flow along all paths between two nodes (u, v). The min-cut section describes the minimum number of edges someone would need to remove from the network to sever all paths between the two nodes. More formally: given two nodes (u, v), the max-flow, min-cut theorem tells you the total capacity for the fewest set of edges you need to remove so there's no path between the two nodes (the cutset). A cut is a graph partitioning $G(S, T)$ such that u is in S and v is in T.

The *capacity constraint* defined in the max-flow, min-cut theorem limits the volume flowing through each edge, per simulation step, to less than or equal to the maximum capacity of the edge:

$$(u \rightarrow v)_{flow} \leq (u \rightarrow v)_{capacity}$$

The *conservation constraint* of the theorem states the amount that flows into each node is equal to the amount flowing out:

$$(u \rightarrow v)_{flow} = \sum_{i=0}^{|\Gamma(v \rightarrow)|} (v \rightarrow \Gamma_{(i)})_{flow}$$

Once again, this can be restated more simply: every node sends out all the resources it receives; it doesn't keep any for itself. This constraint applies to all nodes except u_α and v_Ω. In terms of our simulation, this means that any user who reblogs the message receives as much information as the post contains, and if someone then reblogs the message from them, that person also receives the same amount of information. The source and sink nodes are treated specially due to their position in the flow. The source node is like a faucet capable of adding a certain amount of resource to the network, so nothing flows into the source, only out. In our graph this is synonymous with a user who is likely to receive a lot of reblogs, but isn't likely to reblog a lot themselves (nodes with high out-degree and low in-degree). Conversely, the sink node is like a sponge that absorbs some amount of information from the network without passing any on. Whatever information hits the sink node is absorbed there.

The `nx.minimum_cut` function in Listing 6-16 uses the max-flow, min-cut theorem to determine the minimum cut value between two nodes (u, v) and the partition created by the cut, returned as a tuple (`cut_value, partition`). The `partition` is a tuple of (`reachable, unreachable`) nodes that indicates which nodes would be reachable and unreachable from u. Recall from the previous definition that a cut partitions the graph so that the two nodes are in separate, disconnected components if the cutset is removed.

```
def player_two_turn(G, uq, omega):
❶ cut_value, partition = nx.minimum_cut(G, uq, omega)
  reachable, unreachable = partition
  cutset = set()
❷ for u, nbrs in ((n, G[n]) for n in reachable):
      cutset.update((u, v) for v in nbrs if v in unreachable)
❸ if len(cutset) >= 2:
      cut = choice(list(cutset))
      G.remove_edge(*cut)
❹ elif len(cutset) == 1:
      cut = list(cutset)[0]
      G.remove_edge(*cut)
  return G
```

Listing 6-16: Updating player 2 with more intelligence

We start by computing the cutset between $u(q)$ and v_Ω ❶. To convert the `partition` tuple into the cutset, we loop over the pairs of neighbors in the `reachable` set ❷. For each node in the set, we loop over all their neighbors in the graph. If one of their neighbors is found in the `unreachable` set, then the edge(s) between the node in the `reachable` set and the node in the `unreachable` set belong to the cutset. Once we've processed all the nodes this way, we'll have the list of all edges required to disconnect the two nodes. If there's only one edge in the set, player 2 chooses this edge for removal ❹. Otherwise, player 2 chooses an edge from the `cutset` uniformly at random, again relying on the `choice` function ❸.

This output shows the result of rerunning the simulation with the new strategy for player 2:

```
Average -0.9200
Average -1.0000
Average -1.0000
Average -1.0000
Average -0.6000
Average -0.8400
Average -1.0000
--snip--
Average -1.0000
Average -0.9200
Average -0.9200
Average -1.0000
Average -1.0000
Average -0.3600
Average -0.9200
```

After analyzing the modified player 2 in 6,250 simulations, I got a population mean of –0.8144 and can say with 95 percent confidence that the population mean for simulations with the modified player 2 is between –0.9484 and –0.6803. When you run the code on your machine, you might see slightly different results (remember, we're dealing with a lot of randomness), but the overall trend should remain consistent. It seems that this simple strategy changes the simulation to heavily favor player 2, even with the network growth still favoring player 1.

There's always a chance we're erroneously claiming that we've improved player 2's chances. Since we can't simulate every possible outcome, we can never be 100 percent sure our population means are accurate. How, then, can we be sure this result isn't due to some random fluke? The truth is, we can be sure only up to a certain point. We have to accept that we can't know for certain. This brings up an important point: we need to think about how much risk we're willing to accept of coming to an incorrect conclusion. When you perform an analysis in the wild, there are often real-world consequences for acting on incorrect conclusions. You should pick a confidence level that complements the amount of risk in the event that you're wrong. The higher the risk, the higher confidence you should require. Once you've chosen your desired level of confidence, you can convert it into your t-test threshold by subtracting the desired confidence level from 100. For example, I want to be very certain our result is not a fluke, so I'll set the confidence level to 99 percent, which means we're willing to accept a 1 percent probability of coming to an incorrect conclusion. We can now use this threshold to test our claim that we've improved player 2's chance of winning.

More formally, we can state the hypothesis that changing player 2's logic has created a significant reduction in the population mean ($h_1 = \mu_0 > \mu_1$). The null hypothesis, then, is that the random sample's mean will be equal to or less than that of the modified player ($h_1 = \mu_0 \geq \mu_1$). We can compare the population mean of this set of results using a statistical method known as the *two-sample t-test*. This t-test quantifies the difference between the arithmetic means of the two samples. A common application is to test if a new process or treatment is superior to a current process or treatment. In our case, we'll use it to determine if the difference between the two population means is significant enough to claim that our changed strategy for player 2 has in fact improved their chance of winning.

The proof of concept uses the `scipy.stats.ttest_ind` function to run this test. The result is an object with an attribute named `pvalue`. The p-value quantifies the probability of observing a value as or more extreme than the tested value, assuming the null hypothesis is true. We compare this number against our threshold of 1 percent to determine if we're confident enough in the result to reject the null hypothesis. In this case I've run the test over a dozen times and every time the improved player 2 score is significantly low enough to support the claim that we are 99 percent sure the change we made to player 2 improved their chance of winning. We can visualize the two probability distributions as in Figure 6-4 to see just how big of an impact the change has had.

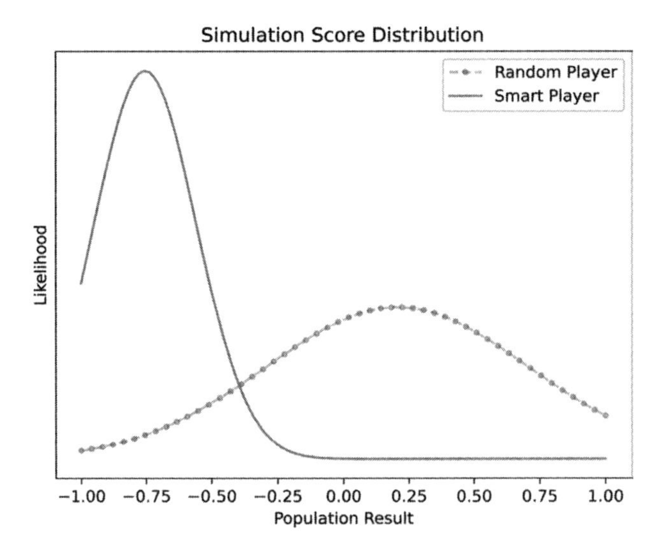

Figure 6-4: Comparing probability distributions

This graph shows the likelihood of all possible test outcomes for both the random and improved player 2 models. The light gray dotted line represents the baseline performance of the random player model. The dark gray continuous line is the performance of the improved player model. The large peak and steep drop-off around –0.8 shows that the improved player performed more consistently and could win most series by a large margin. In fact, it would be incredibly unlikely for any series of tests to average as high as –0.25 using the improved player 2 code. We can interpret this as an indication that the ability to selectively remove edges has the potential to highly disrupt the information flow within this network.

You can run the proof of concept using the command `python mcs_multiplayer.py` in the *Chapter 6* directory in the book's supplemental materials. On each execution, the code runs a group of simulations for both player 2 types, then calculates the population means and compares them using the one-tailed t-test. It will output a line telling us whether or not we can reject the null hypothesis, and finally it generates a graph like the one in Figure 6-4 for analysis. As an exercise, try adjusting the weights in `XI` to more heavily favor new connections and see if this impacts the result to player 1's benefit. What other changes could you make to player 1 to improve their ability to defend the network?

Summary

The concepts introduced in this chapter—Monte Carlo simulations, finite state machines, random walks, weighted choice—combined with the foundational graph theory from the last three chapters make up an extremely flexible set of tools that go far beyond social network analysis. By defining finite state machines for simulations, analyzing repeated simulations to determine the likelihood of a particular outcome, modifying simulations to get different results and insights, and modeling how graphs may evolve over time, you can quantitatively assess security risks by modeling potential changes to the environment.

One scenario where I constantly find myself applying Monte Carlo simulations is in crowd flow dynamics. Predicting how people will move through an area, where they'll gather, and how they might change that movement in response to different types of obstructions is one of the keys to planning effective physical security controls. We'll discuss this a bit more in the context of the art gallery problem in Part III, but you may already have some idea of how you could approach this task using what we've covered here.

This is just the beginning of Monte Carlo simulations, though. By changing the logic at each simulation step, you can model all kinds of unique behaviors in the network. Designing an appropriate simulation is as much an art as it is a science, so don't be afraid to branch out and explore some wild simulation ideas.

To help you as you go forth, the Jupyter notebook that accompanies this chapter has code to display random walks in 2D and 3D, which you can use to visualize the simulations you develop. Often, seeing the results distributed visually can lead to interesting discoveries (like paths that always cross a single point). By combining the random walk display code with the animation code from the supplemental materials, you can even create a video of the simulation.

As you explore the related literature, you'll find numerous advanced discussions of how to select a "best move" within the Monte Carlo simulation. As you saw in the proof of concept, small changes to parameters can have a drastic impact on the result. It's important to be aware of the rationale and implications of each change to the model so that you can formulate more accurate assessments and draw well-founded conclusions about the networks you simulate.

You can learn more about GGP theory and algorithms from Stanford University's online course (*http://ggp.stanford.edu*). Several of these models lend themselves well to various information security tasks, such as risk analysis, budget planning, and incident response. If you'd like to learn more about information flow, check out the research paper "An Information Flow Model for Conflict and Fission in Small Groups,"[11] which describes a formal process for measuring information flow and detecting unbalanced sentiment in a social network.

7

USING GEOMETRY TO IMPROVE SECURITY PRACTICES

When you think about modern security, geometry might not immediately come to mind. However, using computer algorithms to solve classic geometric problems such as line length, shape area, and object intersections allows you to analyze spatial relationships, which in turn can inform your security practices. This field is known as *computational geometry*.

In the next few chapters, we'll apply geometry to problems that relate data to the physical world using geometric rules. In Chapter 8, we'll use the properties of shapes to locate an imaginary perpetrator using cell phone data and MapBox Maps through a process you've probably heard of: location triangulation. Then, in Chapter 9, we'll put on our resource planning hats and look at the distribution of emergency services (like fire stations) using one of my favorite geometric algorithms, Voronoi tessellation. Finally, in Chapter 10, we'll explore using geometry for facial recognition.

But first, we need to cover computational geometry more generally. We'll begin this chapter by going over the essential theory with Python examples. We'll use a Python library called Shapely, which abstracts away a lot of common tasks such as defining shapes and checking if two shapes touch or overlap at any points.

Describing Shapes

Geometry is one of those areas of mathematics that is relatively intuitive for human brains. If I showed you a shape and asked you to label it as a square, triangle, or circle, you wouldn't even have to consciously think about the definition of the shape; you'd simply "know" the matching name. Encoding geometric intuition into computers has proven less straightforward. Before we can discuss how to analyze shapes, then, we need a way to describe the shapes in a data format our programs can digest. Then we need to define a series of mathematical checks for each type of shape we want to be able to identify. Luckily, most of the heavy lifting has already been done for us in Python's Shapely library. In this section, we'll use Shapely to define some of the key shapes we'll need to understand for the upcoming projects to make sense. We'll start with the very basic building blocks of shapes: points and lines. From there we can build increasingly complex shape representations, combine shapes to form 2D models, and look at some of the interesting analysis functions Shapely provides.

Points and Lines

We start with an empty universe represented as a 2D Cartesian plane. Only two types of object exist in this world: a point and a line segment. A *point* denotes an exact place on the plane using the common (x, y) coordinate system. A *line segment* is a part of an infinitely long line; it is bounded by, and contains every point on the line between, two distinct end points. In theory there are an infinite number of points between the two end points; in practice, the number of distinct points is limited by the floating-point precision of the platform or programming language. Python supports 17 decimal places of precision, which will be more than enough for our tasks.

In Shapely, you define a point by telling the library what the point's associated x and y values should be, as in Listing 7-1.

```
from shapely.geometry import Point
point_a = Point(2.0, 4.0)
point_b = Point(0.0, 0.0)
```

Listing 7-1: Defining a point in Shapely

This code creates two Point objects on our Cartesian plane: point_a at (2.0, 4.0) and point_b at (0.0, 0.0).

To create a line segment, we can call the LineString class in Shapely and pass in the starting point (*x*, *y*) and the end point (*x*, *y*). We can pass in either tuples or Point objects. Listing 7-2 shows how to create a LineString object from scratch or from the two previously created points.

```
from shapely.geometry import LineString
line = LineString([(2.0, 4.0), (0.0, 0.0)])
line2 = LineString([point_a, point_b])
```

Listing 7-2: Creating a LineString *from* Point *objects*

The two line segments defined in Listing 7-2 are identical from the library's perspective, so you're free to use whichever syntax suits your code best. If you plan to reuse the same points throughout your application, creating the Point objects first makes the code cleaner and easier to understand.

Points and line segments are the most basic building blocks of shapes. Let's look at how we can combine them to define more complex shapes known as polygons.

Polygons

By combining points and line segments in various configurations, we can build polygons like squares, stars, and even a very close approximation of circles. In geometry, a polygon P is a plane figure that is described by a finite number of points connected by line segments that form a closed chain. A *closed chain* means that the first and last point in the sequence are always the same, so the series of line segments always ends by connecting back to the origin. Line segments used in constructing P may be referred to as *edges*, *sides*, and *faces* interchangeably. The points along the boundary of P are often referred to as the *vertices* (singular *vertex*) of the shape. Listing 7-3 shows how to create a polygon with the Shapely library.

```
from shapely.geometry import Polygon
poly_a = Polygon([(0, 0), (10, 0), (7, 5), (3, 5)])
poly_b = Polygon([point_a, point_b, (10, 0), point_a])
```

Listing 7-3: Creating a Polygon *from* Point *objects*

First, we import the Polygon class, which allows us to define polygons similar to how we defined LineString objects in Listing 7-2. The major difference with the Polygon class is that Shapely creates an additional point to close the shape boundary if it isn't explicitly defined (as with poly_a). The second object, poly_b, shows how you can mix Point objects and hardcoded values when defining a polygon's perimeter. It also demonstrates how to explicitly close a polygon by including a final point that matches the initial point in the perimeter. Remember: whenever possible, explicit is better than implicit.

A *regular* polygon is one where all angles and sides are equal. Any polygon that doesn't satisfy this criteria (which is the majority of them) is considered *irregular*. A *simple* polygon has only one boundary, which doesn't intersect itself at any point. A *complex* polygon has one or more edges that intersect, making the shape twist across itself. Many rules about simple polygons don't work for complex polygons, so we'll avoid them.

Polygons can be very complex shapes with hundreds of edges and vertices, but all possible polygons can be classified into one of two categories. A *convex* polygon is one where all its interior angles are less than 180 degrees. All the vertices of the polygon will point outward, away from the interior of the shape. A *concave* polygon is any polygon that does not satisfy the convex definition. A simple rule of thumb is that if you can travel counterclockwise around the entire perimeter of the shape using only left turns, then the shape is convex. If you ever have to veer right to get to the next vertex, then the shape is concave. Figure 7-1 illustrates the difference.

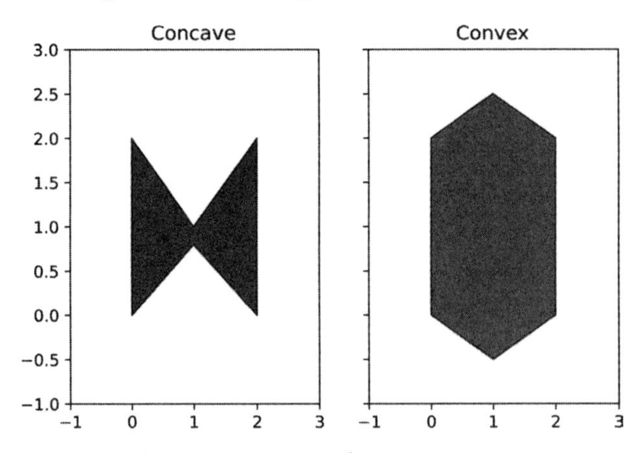

Figure 7-1: Concave vs. convex shapes

NOTE *The code to generate Figure 7-1 is in the 11th cell of the* Computational_Geometry.ipynb *notebook.*

The shape on the left has two points where the perimeter moves toward the interior (1, 1) and (0.8, 1). If you were walking around the perimeter of this shape counterclockwise, you'd have to turn right when you reached either of these points. The shape on the right has no vertices that point inward. You could indeed walk counterclockwise around the whole shape using only left turns, so it is convex.

Similar to how you can combine points and line segments to make basic shapes, you can combine a simple polygon and a special object called a *linear ring* polygon (*ring* or *hole* for short) to create voids in the interior. Figure 7-2 shows the result of combining a simple polygon with a ring.

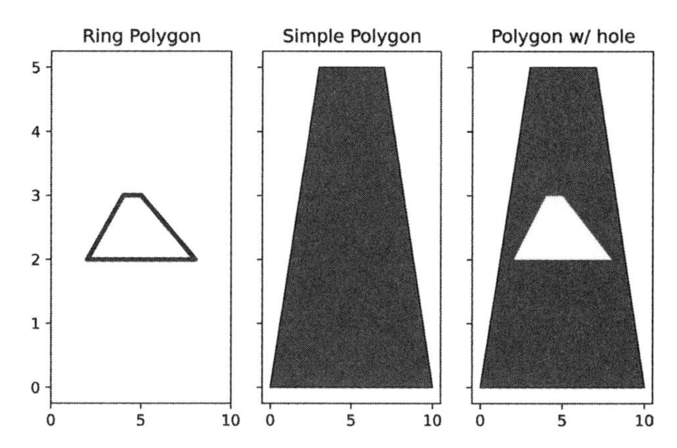

Figure 7-2: Comparing polygon types

NOTE *The code to generate Figure 7-2 is in the 12th cell of the* Computational_Geometry .ipynb *notebook.*

A ring polygon *rph* (the leftmost image of Figure 7-2) is a hollow shape with no solid region. It's composed entirely of the points and line segments that fall along its defining border, the thick black outline in the image.

A simple polygon *P* (the middle image) is a solid plane region that contains all the points within its boundaries as well as all the points along the defining perimeter, shown by the black shape in the middle.

The polygon with a hole (the right image) combines the ring polygon on the left with the simple polygon in the middle. To combine the two, we subtract the points in the ring from the set of points in *P* ($P2 = P1 \notin rph$). The points that fall within the ring shape are excluded from the geometry of the overall polygon.

Listing 7-4 shows how we can construct the three different classes of polygon in the figure.

```
from shapely.geometry import Polygon, LinearRing
poly_a = Polygon([(0, 0), (10, 0), (7, 5), (3, 5)])
poly_hole = LinearRing([(2, 2), (8, 2), (5, 3), (4, 3)])
rbus_b_holed = Polygon(poly_a, [poly_hole])
```

Listing 7-4: Creating a polygon with a hole in Shapely

We define the main polygon, `poly_a`, the same as in Listing 7-3, using its bounding vertices. To define the ring polygon, `poly_hole`, we create a `LinearRing` object by passing in the vertices defining its perimeter. We create the final polygon by combining the two shapes into a new `Polygon` object. The `poly_hole` object is passed inside a list to support multiple holes in the same polygon. Holes may not cross each other or the exterior of the polygon; they can touch only at a single point. Shapely doesn't prevent you from creating invalid features, but it will raise exceptions when you try to operate on them.

By combining individual points, line segments, polygons, and holes, we can begin to model the physical space around us with enough accuracy to be useful. We can use complex polygons to define the bounds of a space and use holes inside the complex polygons to denote areas people can't travel due to obstructions. We'll dive into this more in the "Scenario: Planning Security for a Concert" section, but first let's cover some best practices to help your code run smoothly.

Vertex Order

When you create a shape, the order of the vertices matters. The standard practice is to pass the vertices in counterclockwise order (`ccw` in Shapely) around the perimeter. Having points in a known order makes several operations—like checking if a point falls inside a shape or if a polygon is convex—considerably faster to compute. Shapely has a couple of functions to help with this. In Listing 7-5 we check if an object is in counterclockwise orientation and, if not, translate its point order so that it is.

```
ring = LinearRing([(0,0), (1,1), (1,0)])
print(ring.is_ccw)
new_ring = LinearRing(list(ring.coords)[::-1])
print(new_ring.is_ccw)
```

Listing 7-5: Checking and fixing the order of vertices

We start by creating a `LinearRing` object. You can prove to yourself on a piece of graph paper that the order of the vertices, from left to right, forms a clockwise description of a triangle. We can use Shapely to check if the vertices are in counterclockwise order by printing the Boolean attribute `is_ccw`, which will be `False`. To translate the list into counterclockwise order, we use a Python list inversion (`[::-1]`), which adds the coordinates in reverse order to a new list, then assigns the newly reordered list to a new `LinearRing` object called `new_ring`. Printing `is_ccw` on `new_ring` will now return `True`, confirming the new ordering is indeed counterclockwise.

Now that you can describe shapes in a way the computer can work with, let's look at some common operations you'll repeatedly use when geometrically analyzing security problems. In the next section we'll cover useful algorithms for finding areas, determining overlaps and intersections, and calculating perimeter lengths for irregular shapes.

Scenario: Planning Security for a Concert

The best way to get acquainted with computational geometry theory is to apply it in a scenario, so imagine you're asked to plan the security for a concert being held at a local park. You need to decide how many security personnel to assign to the event and where they'll be positioned, and you need to provide the event coordinators with recommendations for safe attendance capacity. Figure 7-3 shows the park layout from an overhead view.

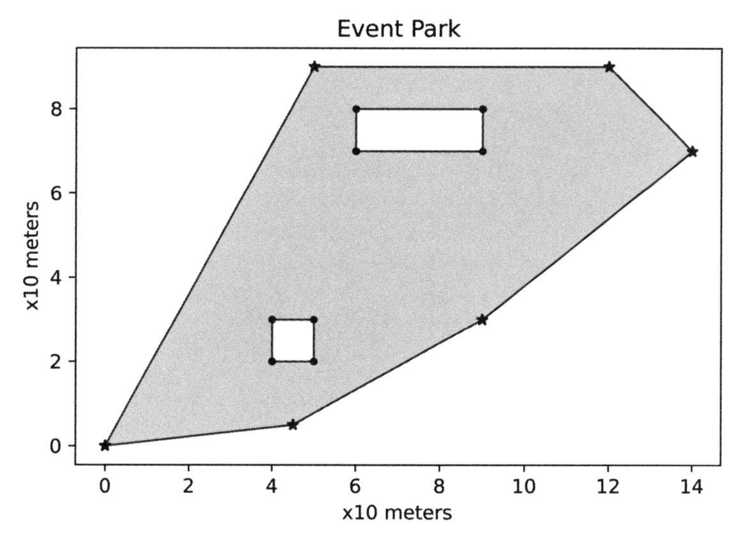

Figure 7-3: The park as a polygon with holes

NOTE *The code to generate Figure 7-3 is in the 13th cell of the* Computational_Geometry .ipynb *notebook.*

The outline of the polygon represents a tall fence around the park. The small square near the bottom of the park is an information booth that can't be accessed by attendees, so we subtract it from the usable area. The rectangle near the top of the park is the stage, which also can't be accessed by attendees, so we remove it as well. The remaining gray area is the part of the park that's accessible to attendees. We'll calculate this area and then use the result to figure out the number of attendees that can safely and comfortably attend the event.

Calculating Safe Occupancy Limits

To find the area of irregular shapes like the park, first we must decompose the shape into a set of simple shapes, like triangles, and then sum the area of each composite shape. The act of decomposing a shape like this is known as *tessellation*, and involves overlaying a plane using one or more geometric shapes, called *tiles*, with no overlaps and no gaps. Figure 7-4 shows two approaches to triangular tessellation for the park space.

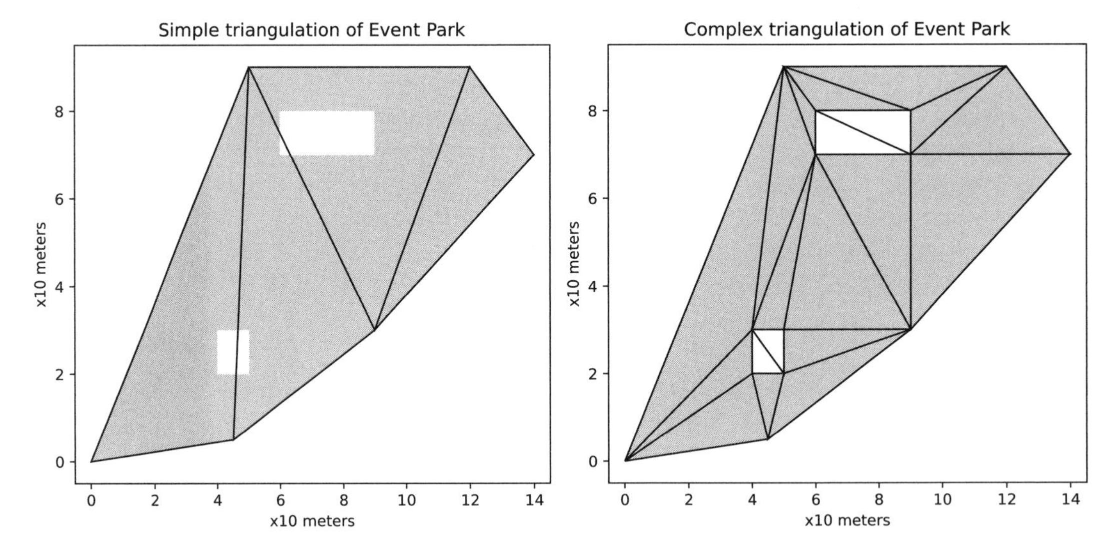

Figure 7-4: A triangular tessellation of a simple polygon and a polygon with holes

NOTE *The code to generate the two plots in Figure 7-4 is in the 16th cell of the* Computational_Geometry.ipynb *notebook.*

The simple approach on the left tessellates the base polygon and temporarily ignores the holes. You calculate the total area of this shape and then subtract the areas for the two restricted zones. I used this method by hand back in my early days. The approach on the right tessellates the complex shape with holes included using Shapely's `triangulate` function (not to be confused with location triangulation, which we'll cover in Chapter 8). We can find the area by summing the n areas of each gray triangle:

$$Area(P_{\Delta}) = \sum_{i=0}^{n-1} \frac{1}{2} (P_{\Delta[i]}B)(P_{\Delta[i]}H)$$

Here $P_{\Delta[i]}B$ and $P_{\Delta[i]}H$ are the base and height measurements, respectively, for the ith triangle in P_{Δ}.

Luckily, we don't have to worry about manual tessellation just to find the area of the complex shape. The Shapely library automatically handles all of this under the hood when you create a shape. Listing 7-6 shows how to create the park shape and calculate the usable area for attendees.

```
park = Polygon([(0,0), (4.5,0.5), (9,3), (14,7), (12,9), (5,9)])
info_booth = LinearRing([(4,2), (5,2), (5,3), (4,3)])
stage = LinearRing([(6,7), (9,7), (9,8), (6,8)])
event_shape = Polygon(list(park.exterior.coords), [info_booth, stage])
event_area = event_shape.area*10
print (f"{event_area} m^2 usable area")
```

Listing 7-6: Creating the complex park shape and calculating the area

We start by creating the outermost Polygon to represent the bounds of the event space. We then create a LinearRing object for both the stage and information booth. Next, we create the complex polygon representing the shape of the usable space in the venue. The library provides an area attribute for polygons that respects any holes we passed in during creation.

When you're dealing with really big shapes like the park, it's common practice to apply a scaling factor—that is, to shrink the whole shape by some known constant to make the numbers easier to work with. In Figures 7-3 and 7-4, one unit is equal to 10 meters, or a scaling factor of 0.1, so multiplying the event_shape.area by 10 adjusts the result for the scaling applied to the map. The output of the print statement is:

```
635.00 m^2 usable area
```

Now that we know the usable area, we can calculate the number of attendees that can safely fill it. Listing 7-7 shows how to convert the area result to attendee count by dividing the usable area by the amount of space we aim to provide each attendee.

```
import math
safe_capacity = int(math.floor(event_area / 0.75)) # 8ft sq
max_capacity = int(math.floor(event_area / 0.37))  # 4ft sq
print("Comfortable capacity: %d people" % safe_capacity)
print("Maximum safe capacity: %d people" % max_capacity)
```

Listing 7-7: Calculating capacity based on usable area

Years ago, I settled on 0.75 square meters—a little over 8 square feet—per person for events with a mixture of seated and standing attendees. (I later discovered others had reached roughly the same numbers for similar scenarios.) We multiply the event_area by 0.75 to find the number of people that can comfortably attend the event with room for some people to sit on a blanket and others to walk around. To find the maximum number of people that can safely stand in an area, we can cut the spacing roughly in half to 0.37 square meters, or about 4 square feet per person. Now, we can multiply the event area by 0.37 to find the "standing room only" capacity of the venue, max_capacity. This tight spacing would be like standing in a crowded hallway: you're not quite bumping into other people yet, but almost. The output from the code should be:

```
Comfortable capacity: 846 people
Maximum safe capacity: 1716 people
```

To estimate the number of security personnel for the event, we'll round the 846 attendees up to 900. From experience, I use a ratio of 60:1 attendees to security, which in this case means 900 / 60 = 15. However, 900 includes everyone that will be in the event area, including security personnel, so in a real scenario you actually want to round in the other direction and recommend a max comfortable capacity of 830 attendees while estimating the security detail for the higher crowd number, 900.

After applying this formula to some real venues, you'll notice your comfortable numbers are usually lower than those recommended by safety codes issued from the fire department, which is intentional. The fire department numbers are concerned with the number of people that can safely evacuate in the event of an emergency and have nothing to do with crowd security or comfort. You'll also find that event coordinators will settle somewhere between the number you suggest and the number the fire department will allow (hence rounding up in planning for personnel and rounding down when recommending attendance).

Determining Placement of Security Personnel

Now that we know how many security personnel we need, let's address where to place them. We'll use tessellation combined with another common operation known as *centroid location*, which finds the point that's equidistant from all of the perimeter vertices of a polygon (that is, the *centroid*). The centroid of a convex object always lies inside the object's area. A concave object might have a centroid that is outside the area, but since we're tessellating the shape first, each resulting triangle will be convex and the centroid will be located within the perimeter.

The centroid of $P = [(x0, y0), (x1, y1), \ldots, (xn-1, yn-1)]$ is the point $C(x, y)$, where

$$C_x = \frac{1}{6A}\sum_{i=0}^{n-1}(x_i + x_{i+1})(x_i\,y_{i+1} - x_{i+1}\,y_i)$$

$$C_y = \frac{1}{6A}\sum_{i=0}^{n-1}(y_i + y_{i+1})(x_i\,y_{i+1} - x_{i+1}\,y_i)$$

The term A is the signed area of P, which is calculated using the *shoelace algorithm*:

$$A = \frac{1}{2}\sum_{i=0}^{n-1}(x_i y_{i+1} - x_{i+1}\,y_i)$$

The points in P need to be in sequential order along the perimeter for this to work. If the order of the vertices is counterclockwise, the area will be a negative value; otherwise, it will be positive. In either case the absolute value of $C_{(x,y)}$ will be correct. We can use this information to place each of the 15 security personnel in the centroid location of the triangle with the largest area that hasn't been assigned a guard yet. Listing 7-8 shows how to achieve this.

```
from shapely.ops import triangulate
tess = triangulate(event_shape)
area_dict = {i: tess[i].area for i in range(len(tess))}
sort_areas = sorted(area_dict.items(), key=lambda x: x[1], reverse=True)
sec_points = [tess[t[0]].centroid.coords[0] for t in sort_areas[:15]]
```

Listing 7-8: A centroid-based personnel dispersion algorithm

We start by triangulating the `event_shape` object from Listing 7-6 using the `triangulate` function. The result is a set of `Polygon` objects representing the triangles making up the shape (denoted as P_Δ). We then use a dictionary comprehension to create a sortable list of triangle index and corresponding area. We use the `sorted` function to sort the `area_dict` by value. Using descending order with `reverse=True` allows us to use the coordinates for the centroid of the first 15 triangles to estimate a good dispersion of security personnel. Figure 7-5 shows the resulting placement plan.

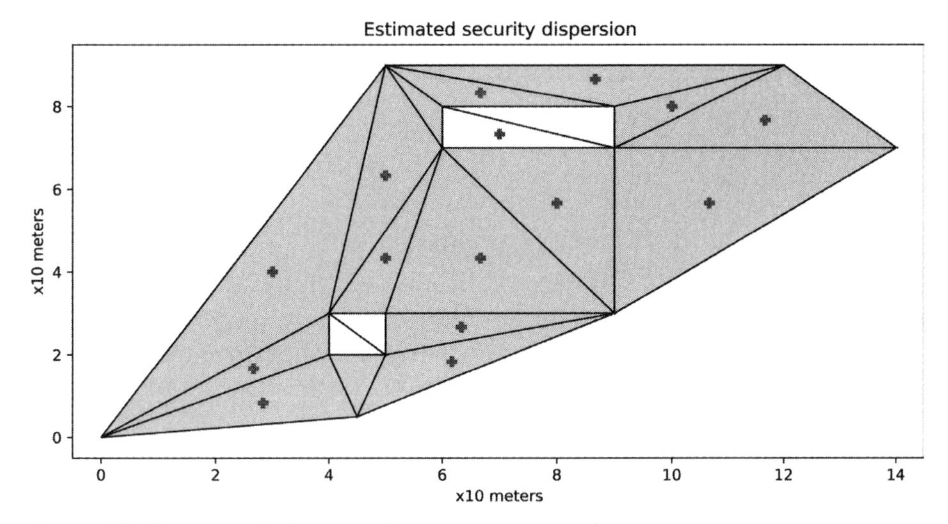

Figure 7-5: The centroid placement for security personnel

NOTE *The code to generate Figure 7-5 is in the 19th cell of the* Computational_Geometry.ipynb *notebook.*

The triangulation is identical to the one shown in Figure 7-4. The gray plus symbols are the centroids for each of the selected triangles. As you can see, the result of this fairly basic algorithm is pretty good. Regardless of where you're standing in the venue, you're never far from one or more security stations. Using a dispersion method like this ensures that security can respond quickly regardless of where they're needed. The one problem is the fact that it placed a guard in one of the unusable parts of the venue, right on stage! We'll see how to improve this algorithm and fix the placement of that guard in the "Improving Guard Placement" section, but for the moment, let's look at another useful piece of analysis that will help us plan and schedule walking patrols.

Estimating Guard Patrol Timing

Perimeter length is a great way to estimate timing for patrols. For simplicity, let's assume the guards patrolling the perimeter of the event don't count against the 15 guards we already placed. We could assign the misplaced guard position to patrol, but that would change the ratio of security to attendees in the venue (albeit only slightly). For now, let's pretend we have another two guards positioned along the perimeter so they don't count toward attendance. One is located at a stationary point (say, at a primary access point like a gate) and another is walking around the perimeter. For safety, the patrol guard should have scheduled check-in times with the stationary guard. The question is how long a normal patrol should take. This is where perimeter length comes in handy. For a polygon, the length of its perimeter is the sum of the lengths of the individual line segments it comprises. To figure out how long the walk should take around the park, we can divide the outer perimeter's length by the guard's estimated walking speed. Listing 7-9 shows how to compute this time.

```
perimeter_len = event_shape.exterior.length*10
walk_time = perimeter_len / (1.1 * 60)
print ("%.2f meter perimeter" % perimeter_len)
print ("%.2f minutes" % walk_time)
```

Listing 7-9: Computing guard patrol times

The exterior `length` attribute of a complex shape like the `event_shape` object is a `LinearRing` object representing the outermost points of the figure. To get the actual distance around the fence, we multiply the `LinearRing`'s length property by the scaling factor. Next we need an estimate for the guard's speed. According to Wikipedia, the average human walking speed at crosswalks is about 1.4 meters per second (m/s), or about 3.1 miles per hour (mph). We can assume our security will be walking a bit slower than this (they're observing the area as they patrol, after all). We set the rate variable to 1.1 m/s or about 2.5 mph. To get the per-minute rate, we multiply the m/s rate by 60. Then, we calculate the `walk_time` with time = distance / rate. The result from Listing 7-9 should be:

```
362.03 meter perimeter
5.49 minutes
```

We can use this information to justify a policy that the patrol guard should check in at a given place along the perimeter every 5.49 minutes. In practice, you'd want to contact the patrol person if you didn't receive a check-in within 6 minutes or so.

Improving Guard Placement

Now let's improve our guard placement from Listing 7-8 by adding the concept of *co-location*, when two objects occupy the same space on the Cartesian plane. With co-location, we'll be able to determine when a guard has been placed in an unusable part of the venue. Shapely provides functions—such as `contains`, `intersects`, `overlaps`, `touches`, and `distance`—to check relationships between geometric objects. Each function takes in a second object as a parameter and answers the question its name suggests. For example, `contains` checks if object B resides completely within the interior of object A. The result will be `True` if no points of B are on the exterior of A and at least one point of the interior of B lies within the interior of A. There's an inverse relationship function, `within`, such that `A.contains(B) = B.within(A)`. The Shapely documentation contains very good explanations of each function.

Listing 7-10 extends the code from Listing 7-8 so that it doesn't place the guard station on the stage.

```
finalized = []
❶ s2 = Polygon(stage)
i2 = Polygon(info_booth)
for guard_station in sec_points:
    i = 0
    new_station = Point(guard_station)
    while any([
      ❷ (s2.contains(new_station)),
        (i2.contains(new_station))
    ]):
        --snip--
      ❸ new_area = sort_areas[15+i]
        i += 1
        poss = tess[new_area[0]].centroid.coords[0]
        if poss not in sec_points:
            new_station = poss
            new_station = Point(poss)
    finalized.append(list(new_station.coords[0]))
```

Listing 7-10: Reassigning guard stations from unusable areas

Because `LinearRing` objects are hollow, they don't actually contain the points within their bounds, so first we need to change the `stage` and `info_booth` objects to `Polygon` objects ❶. A polygon is assumed to be filled, so any point within its bounds will return `True` when we call the `contains` function ❷. By looping over each set of coordinates in the `sec_list`, we can cast them to `Points` and check if the stage or information booth polygons contain that point. If so, we can reassign it to one of the other unoccupied triangles by starting at the 16th triangle (index 15) and assigning the guard station to the new position ❸. We repeat the process of assigning the guard to a new position and checking if that position is contained by one of the unusable areas until we find a usable triangle. Figure 7-6 shows the result of running the improved code.

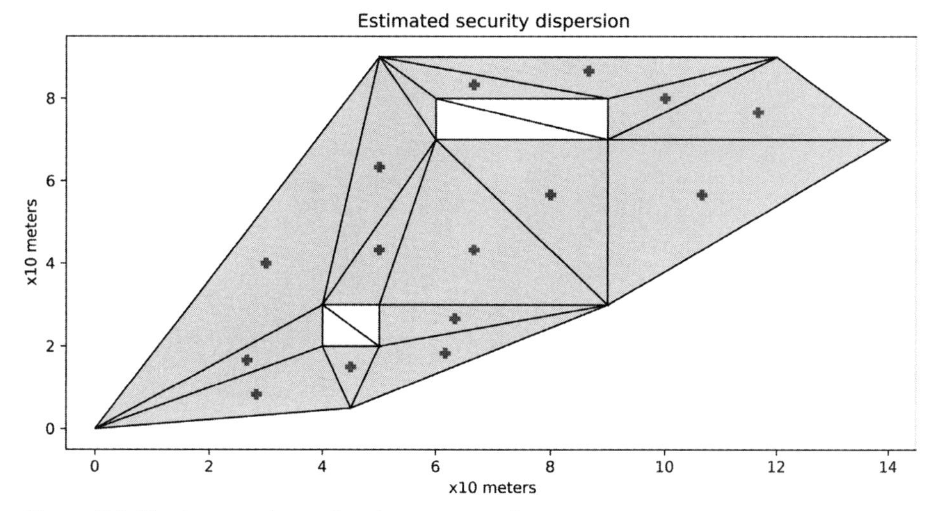

Figure 7-6: The improved security placement results

NOTE *The code to generate Figure 7-6 is in the 21st cell of the* Computational_Geometry.ipynb *notebook.*

When you compare Figure 7-6 with Figure 7-5, you can see the guard station got moved from the stage area to the area below the information booth, which was previously empty. This change places all 15 guards in usable areas of the venue, making sure to assign guards to the largest unprotected sections first. By combining this information with the attendance recommendation and the scheduled patrol routes, we have a good start on an event security plan. We can take this basic recommendation and further refine it with our knowledge of the venue, the event taking place, and so on, to produce a fully formed security strategy that should include contingency plans for more security personnel and attendee overflow areas.

Summary

The power of computational geometry for security lies in the ability to encode physical features in a way machines can interpret: as points, line segments, and polygons. Once you encode the features, you can calculate the area, centroid location, perimeter length, and object relationships to analyze security problems. You've already seen how you can use geometry to describe a physical space like a park and place security personnel effectively. You've also seen how shapes can be combined to make more complex representations using hole objects. Finally, we covered some other common functions that you'll encounter often in practice. There are many more functions available in the Shapely library, but the ones presented in this chapter represent a majority of the work you'll need to perform when analyzing problems geometrically.

As you'll see over the next three chapters, we can combine these operations to create other very interesting security tools. There's still plenty of research to dive into on your own. Geometry has a long history in the security industry, and many research questions can be restated such that you can apply geometric algorithms to them. Of course, none of this even touches on geometry in cryptography, which is its own beast and needs a book of its own. If you're interested in this field, though, check out the paper "Geometric Cryptography: Identification by Angle Trisection."[1] If you're interested in how you can apply geometric algorithms to privacy issues, you'll want to research the field of privacy-preserving computational geometry.

Let's continue our exploration of computational geometry with the topic of geographic location data. In the next chapter, we'll combine MapBox and OpenCellID data to triangulate the location of a cell phone.

8

TRACKING PEOPLE IN PHYSICAL SPACE WITH DIGITAL INFORMATION

Let's continue our discussion of computational geometry with a somewhat controversial topic: tracking people in physical space using digital information. It's no secret that law enforcement agencies across the globe rely heavily on cell phone tracking to locate and apprehend suspects. You might think this requires them to get a warrant and then subpoena the GPS records from a wireless provider, but this isn't always the case, nor is it strictly necessary. Using publicly available information and basic geometry, you can (somewhat) accurately place a phone in a given area, even if it doesn't have GPS. As long as it's connected to a cellular or wireless network, you have a good shot at locating it.

The triangulation process relies on knowing a few things: first, the location—that is, physical latitude and longitude—and configuration for a large number of wireless network hubs, like cell phone towers; second, the approximate broadcast range of these devices; and finally, the signal strength at the cell phone for a set of networks the device can communicate with. This may seem like a tall order, but the first two have already been handled for us by the nice folks at Unwired Labs with their service OpenCellID. The third topic we'll discuss in the next section.

Once you have a basic understanding of the data, we'll cover some of the ethical implications surrounding this project. After that, we'll look at the OpenCellID API in more detail and discuss how to find the location of individual towers and use the API's Geolocation solver to get an address and an accuracy estimate.

Gathering Cellular Network Data

There are several options for getting network information from a phone, ranging from sophisticated hardware attacks to very simple ones that use built-in utilities. Hardware attacks are beyond the scope of this book, but check out the books *Android Security Internals*[1] and *The Android Hacker's Handbook*[2] for an introduction to this topic for Android phones, as well as *iOS Hacker's Handbook*[3] for Apple devices.

The network interface data we need seems so simple and innocuous that it's often made available to almost every application running on a cell phone, tablet, laptop, or other wireless-enabled devices. If you're like me, you're probably very picky about which applications can access your location data directly, but just about every app has a built-in reason to require network access: updates. Unfortunately, this means these applications can see which networks are visible and how strong their signal is. Thanks to the OpenCellID API, this information is almost as good as having a GPS signal, so if you had an exploit for any one of these apps, you could potentially use that to retrieve the information.

As an experiment, you can try to manually scan the cellular networks in your area. Most phones have an option to manually select the cell network you want to connect with. To find it, search your phone's make and model along with "manual cell network selection." There's also a plethora of apps available to perform this task, although I prefer to use the phone's built-in capabilities before downloading any. Suffice to say, from this point forward I'll assume you have access to the device's network data through some legal means. I've also provided the sample data I used when developing the examples so that you can follow along without access to any device.

We'll be working with the OpenCellID API throughout this project, so let's take a moment to discuss what it is and why we'll be using it. The OpenCellID API is an online service that stores information about cellular network base stations, such as location and network type. *Base stations* are what people usually mean when they refer to a cell network's towers. Technically, the part of the tower we're used to seeing just holds the physical antennas; the real brains of the operation are housed near the base of the antennas. These boxes contains the hardware and application logic necessary to route traffic from devices onto the cell provider's network infrastructure and its final destination. For the rest of the chapter, I'll use the terms *base station* and *tower* interchangeably.

After creating a free account and getting an API access token, you'll have access to the data through an online REST API. Having publicly accessible data about cell networks is useful in a number of security applications, aside from tracking physical locations. If you're traveling within any metropolitan area, your phone is being bounced from tower to tower. This usually happens seamlessly in the background and can occur for a couple of different reasons. Cell networks are constantly balancing traffic by asking devices to connect to other towers. When a base station detects that it's becoming bogged down with traffic, it may tell new devices to find a different tower to handle their traffic. Your phone is also programmed to try to connect to the base station with the strongest signal first and then try others in descending order as necessary. Having your device bouncing across all these towers creates the potential for an adversary to introduce a rogue base station (one they control the application logic for) with a strong enough signal to trick nearby devices into connecting to it. As your phone is bouncing around, having access to a publicly curated list of known cell towers can help to ensure your device connects only to legitimate cell networks. The online portal for OpenCellID (*https://opencellid.org*) can also be used to explore the data available in a geographic region.

Figure 8-1 shows the OpenCellID map for the Pike Place Market area of Seattle, retrieved from its online portal.

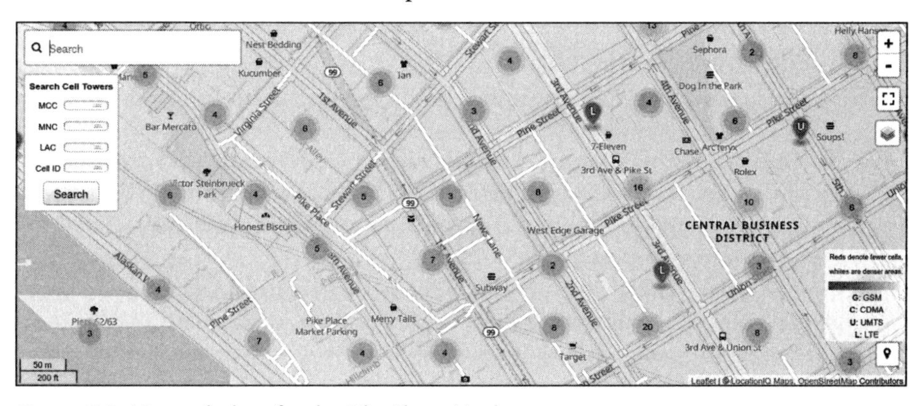

Figure 8-1: Network data for the Pike Place Market area

This image was captured from the OpenCellID main site. The gray circles are clusters of radios that OpenCellID has information for. The number in the center of the circle shows how may radios are in that cluster. You can also see a few pins with letter labels. These are single network instances where the letter denotes the type of radio that's being used to communicate. The key in the lower right of the map shows the four major network types available: GSM, CDMA, UMTS, and LTE. The Jupyter notebook that accompanies this section has more details on each type of network.

These four network types make up the core radio towers we'll use to locate equipment based on their proximity to a set of fixed radio locations. There is another network type that isn't shown on the map, Wi-Fi routers, which you can also use to get more precise results in some cases. The problem is, when compared to cellular network towers, Wi-Fi routers are ephemeral. They may be turned off, or worse (for our purpose), relocated without the OpenCellID database being updated, which would throw our analysis into a tailspin. We'll stick to cellular towers in the example data, but you should definitely explore the Wi-Fi option on your own. There are online services such as WiGLE (*https://www.wigle.net*) that are similar to OpenCellID but for Wi-Fi networks.

Each tower is uniquely identified by a country code and network ID. The data contains estimates on the effective range for the type of network, such as 1 kilometer for most CDMA antennas; it also contains the approximate latitude and longitude of the antenna, other bits to track when it was last verified, and more. By taking several visible network towers and seeing where they overlap service, we should be able to determine the small portion of earth where the device could be located. To do so, we'll create polygons representing the service areas for each base station based on the network type. We'll then overlay the polygons in relation to one another and find the area that all the networks have in common using the Shapely library's `intersection` and `difference` functions.

Before we dive into the nuts and bolts of the data, there are some privacy concerns we should consider. The decisions we make as security researchers can negatively impact the privacy and security of large portions of society in unintended ways. In the next section, we'll cover some considerations to keep in mind before you undertake any project related to device tracking.

Ethics of Tracking Devices and People

In some ways, criminals have it easy. A criminal doesn't have to consider the ethical or moral implications of their actions. Whether they're exposing their target to additional risk is usually not high on their list of concerns. White hats, on the other hand, have ethical, and oftentimes legal, barriers to consider when obtaining or using location information. In the past, I've received records similar to the data for this project from company-owned devices via remote administration tools after an employee's laptop or cell phone went missing. However, even this seemingly benevolent use is an ethical gray area. Here in the US there isn't a clear line on an employee's right to privacy.[4] It's easy to say, "If a company owns a system they have the right to track and monitor it," but what about people who carry their work systems with them during off-hours (something I regularly do myself)? Several companies even require their managers to keep work phones on them during their downtime—even on vacation. There's nothing technical to block these companies from tracking their employees during their private time, and the laws are murky and vague, so it becomes purely a question of ethics. When dealing with corporate entities, being forced to trust their ethical behavior scares a lot of people!

Understanding how this technology works, and how to apply it ethically to improve security without hurting privacy, falls to us as researchers and analysts. After reading this chapter, take some time to look up the relevant laws for your area and, perhaps even more importantly, think about what you believe are appropriate and inappropriate uses for this technology.[5]

For our project, I collected the data on myself using an Android tablet. As the data subject, I was aware of the data collection and gave myself, as the author, permission to use the data in the limited scope of writing this material. The key point here is that the data subject (me) was informed and gave consent. Getting informed consent to perform your analysis can prevent a lot of ethical risk before you even begin. If you apply this type of tracking technology outside the scope of informed consent (such as law enforcement or military applications), you must decide for yourself what ethical doctrine applies.

Now that we've discussed the ethics of device tracking in general, and more specifically concerning the various potential applications of our project, we can get into the meat of the problem. In the next section we'll take a deeper dive into the OpenCellID API, covering the basics of calling the API, the structure of the data returned, and how we can process this information into relevant shape objects.

The OpenCellID API Structure

Technically speaking, OpenCellID is a RESTful API that uses client secret keys to identify users. To access the OpenCellID data, you'll need to sign up for one of these API keys. It's free and easy: you simply provide an email and a use case (like "Research"), and you'll get back an alphanumeric key. Your account comes with a limit of 5,000 requests a day, but this should be more than enough for most applications. If you're smart about caching responses, you should be able to spread those out even more.

There are two common workflows supported by the API. *Geolocation* is the process of turning latitude and longitude information into a location on earth, like "123 Main Street, Seattle, WA." Geolocation comes up a lot both inside and outside of security, so it's a good idea to become familiar with the process. *Geocoding* works in the opposite direction: you take an address and return the latitude and longitude of that point. We'll mostly be working with the Geolocation portion of the API, but it's worth noting OpenCellID also has API calls to help with map displays and calls for monitoring your usage. As you develop your own application, you'll want to take advantage of those additional features.

Listing 8-1 shows the structure of a simple request to the API.

```
payload = {
  ❶ "token": "alphanumeric_code",
  ❷ "radio": "cdma",
    "mnc": 120,
  ❸ "mcc": 310,
  ❹ "cells": [{
      ❺ "lac": 23319,
        "cid": 192337670
    }],
  ❻ "address": 1
}
```

Listing 8-1: The structure of the API payload

Every request needs to contain the alphanumeric `token` that you received when you signed up ❶. The `radio` field ❷ identifies the primary network type of the device we're looking into. Setting this field doesn't restrict the types of radios that you can pass in the `cells` field ❹, however, which represents the base stations that are visible to the device. We'll revisit the `cells` field more in a moment, but first, let's discuss how networks are identified and grouped.

Each country is assigned a three-digit *mobile country code (MCC)*. Most countries actually have several MCCs assigned to break up the geography into smaller chunks that make it easier to manage the traffic for any one region. To uniquely identify a mobile subscriber's network, the MCC is combined with a *mobile network code (MNC)* into the *home network identity (HNI)*, which concatenates both pieces of information into one string.

In Listing 8-1, the device has an MCC of 310 ❸ (which is the first of seven MCCs assigned to North America, numbered 310–316) and an MNC of 120, so the HNI is 310120. Using the HNI, OpenCellID can determine what service provider and network segment a device belongs to when it sends a query to the API. You can pull this information off the device you're testing or pass some default network MCC and MNC, which is what we do in the code for this project. If you look up the HNI from Listing 8-1, you'll see it belongs to the Sprint Spectrum network (*https://imsiadmin.com/assignments/hni*) in North America.

Now, returning to the `cells` field, we see a list of nested JSON objects. We can send between one and seven radio identifiers to OpenCellID to help pinpoint location more accurately. You can even contact Unwired's development team to increase that number if you need to (but that's unlikely).

The next section of the data contains the radios we'd like to retrieve information about. The radio objects we send can be any mix of supported radio types. We identify each radio using two numbers based on its physical locations. Cell networks are divided into several geographic areas, each of which can support between 1 and 65,534 base stations. Each geographic area is assigned a unique *location area code (LAC)*. Similar to the area code prefix on a telephone number, a LAC describes roughly where the base station is located. The second number is the *cell ID (CID)*, which identifies each individual base station within a LAC ❺. You can think of the LAC and the CID like a zip code and a street address, respectively. They work together to create a unique identifier for every base station on a network. Finally, the `address` field ❻ tells the API to return the human-readable address along with the other result fields. If you don't need the address, you can save some bytes by excluding this field.

Listing 8-2 defines a Python function called `lookup_tower` to send the single radio payload from Listing 8-1 to the API.

```
def lookup_tower(payload):
    url = "https://us1.unwiredlabs.com/v2/process.php"
    response = requests.request("POST", url, data=json.dumps(payload))
        return json.loads(response.text)
```

Listing 8-2: A function to call the API and decode the response

Before running anything, let's begin by verifying that the URL presented here (`https://us1.unwiredlabs.com/v2/process.php`) is still the most up to date and appropriate for our use case. Unwired hosts several API endpoints around the world, so there may be another that is closer to you or one with less traffic (which helps reduce latency). You can do this by going to the Unwired Labs API list at *https://unwiredlabs.com/api* and selecting Endpoints from the options on the left. You can then copy the URL for the API endpoint that is geographically closest to you.

We must convert the JSON payload into a string object before passing it to the request library. To do so, we call the `json.dumps` (short for *dump string*) function on our JSON payload. We pass the function to the request library through the `data` parameter. The `response` object we get back will be a JSON object transmitted as text unless there's an error, in which case we'll get a nasty blob of HTML and the function will choke when trying to decode the `response.text` property. To avoid this, we'd expand our production code to wrap the call to `requests.request` in a `try...catch` block or other safety net.

Listing 8-3 shows the format of the response.

```
{
    "status":string,
    "balance":int,
    "lat":float,
    "lon":float,
    "accuracy":int,
    "address":string
}
```

Listing 8-3: A single-tower JSON query response

The status string will be `"ok"` in cases where the request succeeded and `"error"` if the API detected a problem. The `balance` field holds the number of requests remaining for the day. We can take the `lat` (latitude), `lon` (longitude), and `accuracy` fields to plot the tower on a map. Finally, if we pass the `address=1` parameter, the `address` field holds this information as a string. Figure 8-2 shows the result of Listing 8-3 plotted on a map.

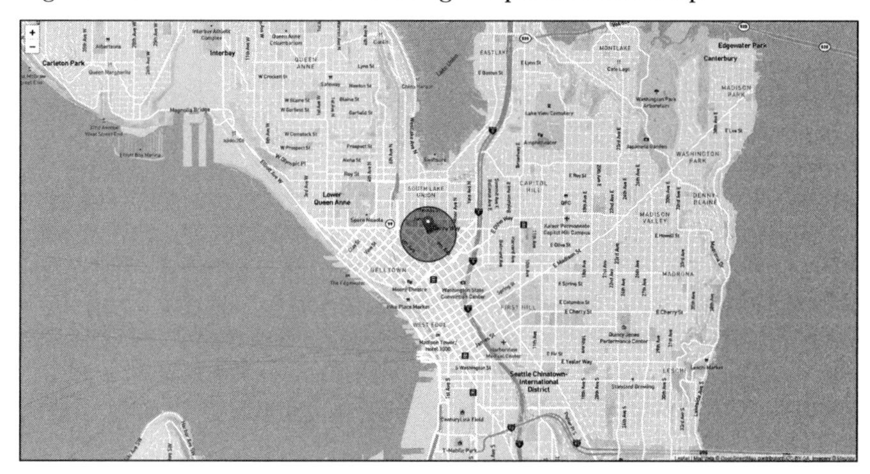

Figure 8-2: Displaying a tower range

NOTE *The code to generate the map overlay in Figure 8-2 is in the 4th cell of the* OpenCell_API_Examples.ipynb *notebook.*

The map area shows the city of Seattle, where the tower is located. The dark gray circle in the middle of the image shows the approximate area covered by the tower, and the pin in the center of the gray circle represents the exact latitude and longitude of the tower. Each of the white lines crisscrossing the image represents a city street. You can see the maximum coverage area of one tower is pretty large—several city blocks at least. If we had only this one tower to try to locate the device, we'd have a lot of ground to cover! In the next section we'll see how to narrow the area further by geolocating several towers and finding their overlapping service region using the GeoPandas and Shapely libraries.

The Proof of Concept: Locating a Device from Nearby Cell Towers

We've seen how we can interact with the OpenCellID API to get back information on towers in our data. Now it's time to put this knowledge together into an application that will locate a device based on the cell towers within its range. We'll extend the `lookup_tower` function from Listing 8-2 to locate each tower in the list recovered from the test device. The data we'll be using to test our application can be found in the file *cellular_networks.json* in the chapter's supplemental materials.

Figure 8-3 shows the four towers from the sample data laid out over a map.

Figure 8-3: Tower signal overlap

NOTE *You'll find code to generate an image similar to Figure 8-3 in the 4th cell of the* OpenCell_API_Examples.ipynb *notebook, but there's a problem with it: the coverage area doesn't scale with the zoom level. Figure 8-3 was actually generated using the MapBox JavaScript API and a CherryPy web server. The code for that server is in the* locator_server.py *file in the chapter's supplemental materials. After starting the server with the Python interpreter, like so*

```
$ python locator_server.py
```

you should be able to navigate to http://127.0.0.1:8080/towers *and see the map as it's presented here. If you get an error like* `"Import Error: Module Not Found: locate"` *you'll need to add the directory with the server scripts to your* `PYTHONPATH` *environment variable. There are many good online tutorials on how to accomplish this for each operating system.*

Your eye can easily locate the area where all the signals overlap, but it isn't quite as easy for a computer to determine. Therefore, our goal is to programmatically identify that portion of the map where all four towers overlap using Shapely to produce a bounded search area. We'll then compare our results with the location guess returned from OpenCellID API to see which is more accurate.

Gathering Tower Locations

Our first step, shown in Listing 8-4, is to gather the location information from each of the four towers using the `lookup_tower` function from Listing 8-2.

```
with open("cellular_networks.json") as f:
    cells = json.load(f)["cells"]
tower_locs = []
❶ for c in cells:
  ❷ payload["cells"] = c
  ❸ tower_loc = lookup_tower(payload)
  ❹ tower_locs.append(tower_loc)
```

Listing 8-4: Gathering tower geolocation information

After loading in the data, we create a `for` loop to loop over each cell ❶. We need to send in each tower by itself to get back its geolocation information; otherwise, we'll get back a location guess using the towers you sent to the API as the reference information. We'll replace the cells in the payload from Listing 8-1 with the tower info loaded in from the *networks.json* file ❷. In a production application, this would be the portion of information you recover from the device you want to track. Next, we call the `lookup_tower` function ❸ and store the result into a list called `tower_locs` ❹.

Now we have a list containing JSON objects, which in turn contain geolocation information in the form of latitude and longitude coordinates for each tower. This is very close to a standard object format called GeoJSON, which many programs of different languages know how to interpret. To make our data more flexible and standardized, let's finish converting the data to GeoJSON format. We can do this simply enough using the pandas data science library and its sister, GeoPandas, which adds support for geometric coordinates and operations. Listing 8-5 shows how to convert the JSON data the rest of the way to the GeoJSON format.

```
import pandas as pd
import geopandas as gpd

tower_df = pd.DataFrame(tower_locs)
tower_df.drop(["status", "balance"], axis=1, inplace=True)
geo_df = gpd.GeoDataFrame(
    tower_df,
    geometry=gpd.points_from_xy(tower_df.lat, tower_df.lon)
)
```

Listing 8-5: Creating a `GeoDataFrame` from tower locations

First, we cast the list of JSON objects to a traditional pandas `DataFrame`, which gives us a chance to clean up any unnecessary fields in the next line. We're dropping the `status` and `balance` fields since they don't add any information to the analysis. Next, we use the GeoPandas `GeoDataFrame` class to cast `tower_df` into the more appropriate geolocation data, which contains a special field, aptly named `geometry`, to hold the geometric representation of each row in the `DataFrame`. In this case, we use the `lat` and `lon` columns from `tower_df` to define `Point` objects, which then get stored in the `geometry` column. The GeoPandas function `points_from_xy` takes in an x- and y-coordinate and returns a `Point` object that GeoPandas can use to associate a shape object with the data. You can see the structure of the `geo_df` data in the output of the 2nd cell in the *OpenCell_API_Examples.ipynb* notebook.

Translating a Geographic Point to a Polygon

We'll be converting these `Point` objects to polygons representing circles, but Shapely isn't aware of specific coordinate systems or units, so first we need to convert the latitude and longitude into native (x, y) coordinates and back again. This requires a rather tricky bit of code that can store intermediate results and keep track of units. It relies on the pyproj (short for Python Projection) and functools libraries to do so. The functools library is for *higher-order functions*, functions that act on or return other functions. Examples include functions that modify the call structure or translate inefficient call flows into more modern and efficient ones. Consider the following example:

```
# A normal function
def complex(a, b, c, d, x):
    print(f"you sent in {a},{b},{c},{d},{x}")
```

The `complex` function takes in five required parameters and prints them all out using a format string. But what if we know that in our use case we'll always call complex with the same first four parameters and x is the only parameter we'll ever need to change? In these situations we can use partial function copies to simplify the calling conventions. *Partial functions* allow us to fix a certain number of arguments for a function and generate a new function that we can call without including those fixed arguments. Here, functools.partial allows us to create a new simplified version of the complex function like so:

```
import functools
# A partial function that simplifies the previous one
simple = functools.partial(complex, 1, 1, 2, 3)
simple(5)
```

The simple function now contains a copy of the complex function, with the first four parameters statically defined to be the values 1, 1, 2, and 3, respectively. Now calling simple with any value will result in that value being passed to the complex function in the x parameter. Calling the simple function with the value 5 is now equivalent to calling the complex function with the values 1, 1, 2, 3, and 5, as we can see from the code's output:

```
you sent in 1,1,2,3,5
```

We'll use the functools.partial function to programmatically create two new functions to handle the coordinate translations. To do so, we'll be using the pyproj library, which is designed to translate between different coordinate systems internally. The Proj class can convert from geographic (lat, lon) to native map projection (*x*, *y*) coordinates and vice versa, which is perfect for our needs. Listing 8-6 shows the function I found and modified from a related GIS Stack Exchange post.[6]

```
import pyproj
from functools import partial
from shapely.geometry import Point
from shapely.ops import transform

def get_shapely_circle(x):
    lat = x["lat"]
    lon = x["lon"]
    radius = x["accuracy"]
  ❶ az_proj = "+proj=aeqd +R=6371000 +units=m +lat_0={} +lon_0={}"
  ❷ wgs84_to_aeqd = partial(
        pyproj.transform,
        pyproj.Proj("+proj=longlat +datum=WGS84 +no_defs"),
        pyproj.Proj(az_proj.format(lat, lon))
    )
  ❸ aeqd_to_wgs84 = partial(
        pyproj.transform,
        pyproj.Proj(az_proj),
        pyproj.Proj("+proj=longlat +datum=WGS84 +no_defs"),
    )
  ❹ center = Point(float(lon), float(lat))
  ❺ point_t = transform(wgs84_to_aeqd, center)
```

```
❻ buffer = point_t.buffer(radius)
  # Get the polygon with lat lon coordinates.
❼ circle_poly = transform(aeqd_to_wgs84, buffer)
  return circle_poly

❽ geo_df["geometry"] = geo_df.apply(get_shapely_circle, axis=1)
```

Listing 8-6: Translating a geographic point to a geographic polygon

We define the `az_proj` (short for *azimuth projection*) string to contain all the variables that will be passed in to the projection code ❶. The most important are `+proj`, which tells the library to convert the coordinates using a method known as *azimuthal equidistance (AEQD)*; `+R`, which holds the radius (in meters) of earth; and `+units`, which tells the code that this number is in meters, but more generally tells the library what units to convert to. The only two variables we need to be able to change on each call are `lat_0` and `lon_0`, which define the (0, 0) point in the coordinate field. Currently our data is in *world geodesic system (WGS)* coordinates. WGS 84 is the standard US Department of Defense definition of a global reference system for geospatial information and is the reference system for GPS (the Global Positioning System). It is compatible with the International Terrestrial Reference System (ITRS) if that's more your cup of tea.[7]

In the `wgs84_to_aeqd` and `aeqd_to_wgs84` functions, we create a partial copy of the `pyproj.transform` function ❷. The `functools.partial` function freezes the first two parameters of the `pyproj.transform` function. Notice in the `aeqd_to_wgs84` function the two calls to the `Proj` class are reversed. That's because the first two parameters to the `pyproj.transform` function define the current and desired representation, respectively. Reversing these two inputs reverses the translation direction, in this case from native coordinates back into geographic coordinates ❸. Since we're creating simplified versions of the `transform` function, we freeze these two parameters in the appropriate order for our translation direction. The remaining parameters to the `pyproj.transform` function define the *x* and *y* values to translate. We'll leave these unfrozen and pass them in when we call the functions in a moment.

Now that we have the two conversion functions defined, we create a `Point` object from the latitude and longitude ❹. Shapely expects these in the reverse order than you might expect. If you try to pass the coordinates directly via the `geometry` column, you'll run into the error `"latitude or longitude exceeded limits."` The `shapely.transform` function (not to be confused with the `pyproj.transform` function) applies a user-defined function to all the coordinates of a Shapely object and returns a new geometric object of the same type from the transformed coordinates. We'll use the `shapely.transform` function to transform the point into native coordinates using the previously defined `wgs84_to_aeqd` function ❺. Shapely points have a `buffer` function that adds a defined amount of space around the point. Essentially, given a point and a desired amount of buffer space around that point, Shapely generates a new set of points representing the location of the boundary of the buffer space. We can use this to generate a circle representing the approximate coverage area of each tower. Since we defined the units in the AEQD transformation function as meters, we can pass the radius for the buffer area in meters as well. This is handy because the `accuracy` field returned by the OpenCellID API is also in meters. The value in the accuracy field describes how much error is in the latitude and longitude. Calling `buffer` with the radius set to the accuracy of the radio creates a polygon that represents the estimated location of the tower.

The signal coverage area is a little more difficult to calculate accurately. If we wanted to be very specific, we could calculate wave propagation, but then we'd need to know the type of tower, its power ratings, its height, and any major obstructions. It turns out that the accuracy field also makes a handy signal strength estimate. The accuracy of the tower placement tends to be about 30 to 50 percent of the optimal signal coverage for the given type of tower. Assuming your target is in a metropolitan area, 30 to 50 percent is also a fair guess for signal dampening, especially without knowing any more about the towers or landscape itself. So I like to use the accuracy field as a quick-and-dirty guess at coverage area as well ❻. By passing the accuracy as the radius to the `buffer` function, we're defining a circle whose bounds will represent the probable coverage area of the tower. In reality the tower may not be in the center of the circle, but it will fall somewhere within it. This means the actual coverage area may be a bit bigger or smaller depending on the landscape and architectural construction of the area around the tower, but this will make a good starting point for our proof of concept.

At this point, we have a circular polygon defined around each tower's *x* and *y* location that represents an approximation of its covered service area, but the points representing the hull of the polygon are currently in native map coordinates. To transform them into geodesic coordinates, we call `shapely.transform` again, this time with the `aeqd_to_wgs84` function ❼, and return the results.

Lastly, we call the `apply` function to get the result of the `get_shapely_circle` function for each row in `geo_df` (setting `axis=1` operates on rows instead of columns) and use the results to overwrite the original `geometry` column in the data with the new polygons ❽.

Calculating the Search Area

In the previous section we tackled the first big hurdle to our project. We can now convert the latitude and longitude of a tower, along with the accuracy estimation, into a geometric object representing the potential service area for that tower. We've also converted the points around the hull of the geometry back into latitude and longitude coordinates we can use with maps. Our next step is to find the geographic area where all of these polygons overlap, or more formally, $A \cap (B, C, \ldots, N)$ where *A*, *B*, *C*, and so forth represent the polygons created during the previous steps. To accomplish this, we'll borrow some code from Stack Overflow[8] that performs repeated Boolean operations to find the intersection between *A* and each other polygon. Let's start by defining a function to handle the simplest case: returning the difference and intersection of two polygons, *A* and *B*.

Listing 8-7 shows the partitioning code.

```
EMPTY = GeometryCollection()
def partition(poly_a, poly_b):
    if not poly_a.intersects(poly_b):
        return poly_a, poly_b, EMPTY
    only_a = poly_a.difference(poly_b)
    only_b = poly_b.difference(poly_a)
    inter = poly_a.intersection(poly_b)
    return only_a, only_b, inter
```

Listing 8-7: Partitioning polygons into difference and intersection elements

First, we check for the simplest case: when polygon *A* doesn't intersect polygon *B* anywhere. In this case, we simply return both polygons along with an empty `GeometryCollection` object. If there's some overlap between the two polygons, we want to return three things. First, we want to return the two differences—that is, the part of *A* that doesn't overlap *B* and vice versa. The two differences are stored in `only_a` and `only_b`. Then, we also want to return the intersection of the two, which can be found using Shapely's `intersection` function.

The code in Listing 8-7 will be used within the main function that solves the intersection code, which uses a slightly modified version of a *sweep line algorithm*, a very famous way to efficiently process an arbitrarily large set of shapes with some Boolean operations (such as unions and intersections). Rather than stopping at each point, we sweep over whole polygons and compare them to all previously known polygons. Each pair of polygons and its subgeometries will be iteratively collected and compared to see which parts of *A* overlap with the other polygons. These overlaps will be treated as subsets of geometry and checked in turn.

Listing 8-8 shows the main function.

```
def cascaded_intersections(poly1, lst_poly):
❶   result = [(lst_poly[0], (0,))]
    for i, poly in enumerate(lst_poly[1:], start=1):
        current = []
        while result:
            r_geo, res_idxs = result.pop(0)
          ❷ only_res, only_poly, inter = partition(r_geo, poly)
          ❸ for geo, idxs in ((only_res, res_idxs), (inter, res_idxs + (i,))):
                if not geo.is_empty:
                    current.append((geo, idxs))
      ❹ curr_union = cascaded_union([elt[0] for elt in current])
        only_poly = poly.difference(curr_union)
      ❺ if not only_poly.is_empty:
            current.append((only_poly, (i,)))
        result = current
      ❻ for r in range(len(result)-1, -1, -1):
            geo, idxs = result[r]
          ❼ if poly1.intersects(geo):
                inter = poly1.intersection(geo)
                result[r] = (inter, idxs)
```

```
            else:
                del result[r]
    only_poly1 = poly1.difference(cascaded_union([elt[0] for elt in result]))
  ❽ only_poly1 = eliminate_small_areas(only_poly1, 1e-16*poly1.area)
    if not only_poly1.is_empty:
        result.append((only_poly1, None))
    return [r[0] for r in result]

❾ polys = list(geo_df["geometry"])
❿ results = cascaded_intersections(polys[0], polys[1:])
```

Listing 8-8: Sweep line algorithm for cascading intersection of polygons[9]

First we create the result field with the first polygon in the list to check for intersection ❶. We loop over the rest of the polygon list and use the partition function from Listing 8-7 to generate all of the intersections and differences for the polygons (*B*, . . . , *N*) ❷. For each of these, we check the geometry to ensure no empty geometry objects get passed through ❸. Once we create all these subgeometries, we can take the cascading union to create the remainder section of polygon *A*. This represents the shape of *A* that doesn't intersect any other polygon ❹. If it is non-empty, we add it to the current list of results ❺.

Next, we once again loop over the resulting intersections ❻ to see which of them also intersects with the primary polygon *A* ❼. We repeat this process until no more intersections are left to check. Sometimes the intersection operation creates tiny polygons, which are really just artifacts that we can toss. To do so, we have a second function that compares each intersection polygon's area to the area of the first polygon; if it's less than $1e - 16 \times$ `A.area`, the polygon is removed. The remaining polygons are assigned back to the `only_poly1` variable ❽ (we'll cover the `eliminate_small_areas` function momentarily). Lastly, we check if the remaining polygon list is empty. If not, we append it to the list of results being held in the `result` variable.

We can now call the `cascaded_intersections` function with the list of tower service areas represented by the shape data stored in the `geometry` column of the `geo_df` data. We create a list containing the shape data we generated in Listing 8-6 representing the tower service areas and assign it to the `polys` variable ❾. We pass the zeroth polygon in the `polys` list as the first argument to the `cascaded_intersections` function. This will be the primary polygon (polygon *A*) the rest of the algorithm is considering intersections for. We pass the remainder of the list as the second argument (polygons *B*–*N*) to tell the `cascaded_intersections` function these are the polygons that may intersect polygon *A*. The `cascaded_intersections` function returns a list of different geometries of interest, which we assign to the `results` variable ❿.

The zeroth element of `results` will be the remainder of polygon *A* that doesn't intersect with any other polygon. The first element will be the intersection of polygon *A* with all the other polygons. The remaining elements will depend on the layout of the polygons but will follow the pattern ($A \bigcap B \notin CD$, $A \bigcap BC \notin D$. . .). We need only the first element, the intersection of all the polygons, but the other results are there for you to explore on your own as well. One way we might use this output is to print out the minimum and maximum values for the latitude and longitude of the intersection result. This gives us a boundary in the shape of a box that completely encloses the geometry of the intersection polygon. We can find this search area pretty simply:

```
x,y = results[1].exterior.xy
print(f"""Search bounded area:
({min(y)}, {min(x)})
to
({max(y)}, {max(x)})""")
```

First we create two variables, x and y, to hold their respective list of coordinate values. Remember that we've already converted the coordinates to latitude and longitude, so all we need to do now is print out the minimum and maximum values from each list to find the latitude and longitude bounds of our search area. The output of the code for our test towers is shown here:

```
Search bounded area:
(47.61858939197041, -122.35438376445335)
to
(47.6221396080296, -122.34381687278278)
```

The coordinates in the output represent the lower-left and upper-right corners and can be used to form a box around the polygon resulting from the cascaded intersection function. We could give this information to a ground team who could go to the area and perform a search.

The code in Listing 8-8 relies on the eliminate_small_areas function, which, by comparison, is very simple to grok. Listing 8-9 shows the code to eliminate any potential artifact polygons.

```
def eliminate_small_areas(poly, small_area):
  ❶ if isinstance(poly, Polygon):
        if poly.area < small_area:
            return EMPTY
        else:
            return poly
  ❷ assert isinstance(poly, MultiPolygon)
  ❸ l = [p for p in poly if p.area > small_area]
  ❹ if len(l) == 0:
        return EMPTY
  ❺ if len(l) == 1:
        return l[0]
  ❻ return MultiPolygon(l)
```

Listing 8-9: Removing small area polygons

First we use `isinstance` to check whether the polygon passed in is an instance of a single polygon ❶. If it is, we check if the polygon's area is smaller than the `small_area` parameter. If so, we return an empty `GeometryCollection`; otherwise, we return the polygon instance. If the object passed in the `poly` parameter isn't an instance of a single polygon, we assert it must then be an instance of a `MultiPolygon` (essentially a list of polygons). If the assertion fails (say you pass in a dictionary by mistake), the code will raise an exception ❷. In the case where the object is a `MultiPolygon`, we use list comprehension to check each individual polygon's area against the `small_area` parameter ❸. If the length of the resulting list is 0 ❹, no polygons remain after the artifacts are removed, so we return `EMPTY`. If exactly one polygon object remains in the list ❺, we return it as a single polygon instance; there's no need to carry the additional weight of a `MultiPolygon` object forward. Otherwise, if more than one polygon remains in the list, we return them all as a `MultiPolygon` object ❻.

After we call the `cascading_intersections` function, we can plot the first item in the result to see the search area identified. Figure 8-4 shows the polygon representing the intersection of all the towers.

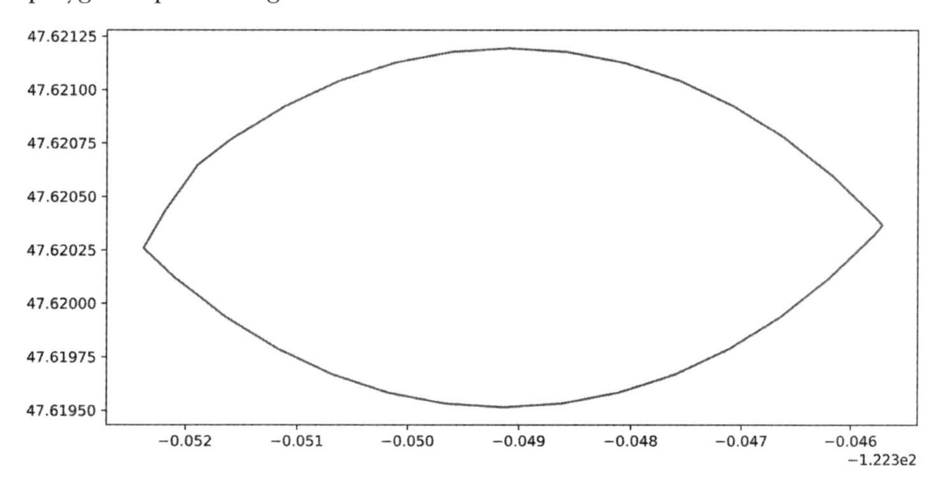

Figure 8-4: Intersection of all four towers as a polygon

NOTE *The code to generate Figure 8-4 is in the 8th cell of the* OpenCell_API_Examples .ipynb *notebook.*

If you compare the shape of the polygon in Figure 8-4 to the overlapping areas from Figure 8-3, you'll see that they look very similar, meaning we've achieved our goal of programmatically identifying the area of interest. We could pass these coordinates in raw form to any GPS device to create a more accurate bounded search area.

Mapping the Search Area for Investigators

We could also overlay the result on a map to see the search area we should pass on to investigators, as shown in Figure 8-5.

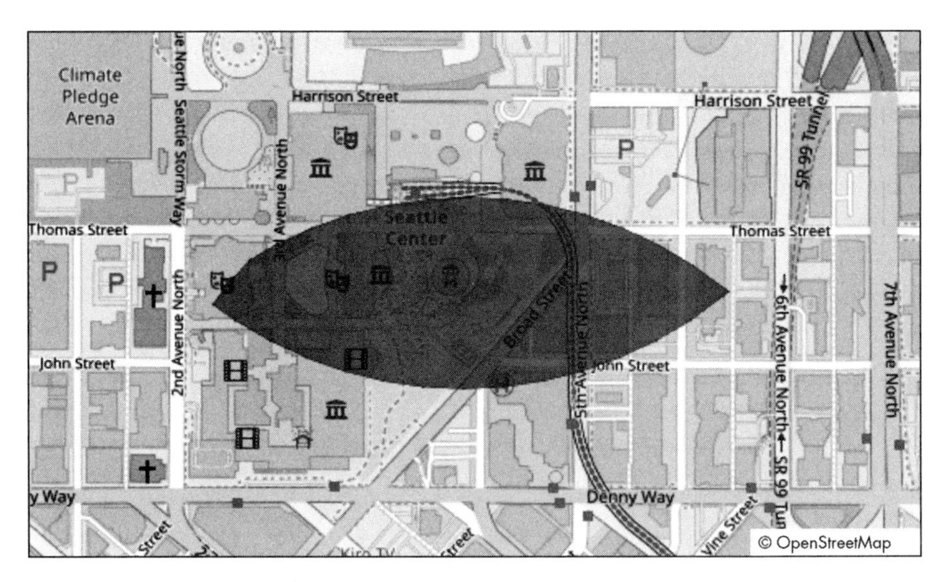

Figure 8-5: Resulting search area

NOTE *The code to generate the overlay in Figure 8-5 is in the 9th cell of the* OpenCell_API_Examples.ipynb *notebook.*

The polygon lays directly over the area known as Seattle Center, home of the Space Needle. Indeed, when I captured the sample data I was standing near the foot of the Space Needle, close to the center of the search area. Now let's compare our result to the one presented by the OpenCellID API, shown in Figure 8-6.

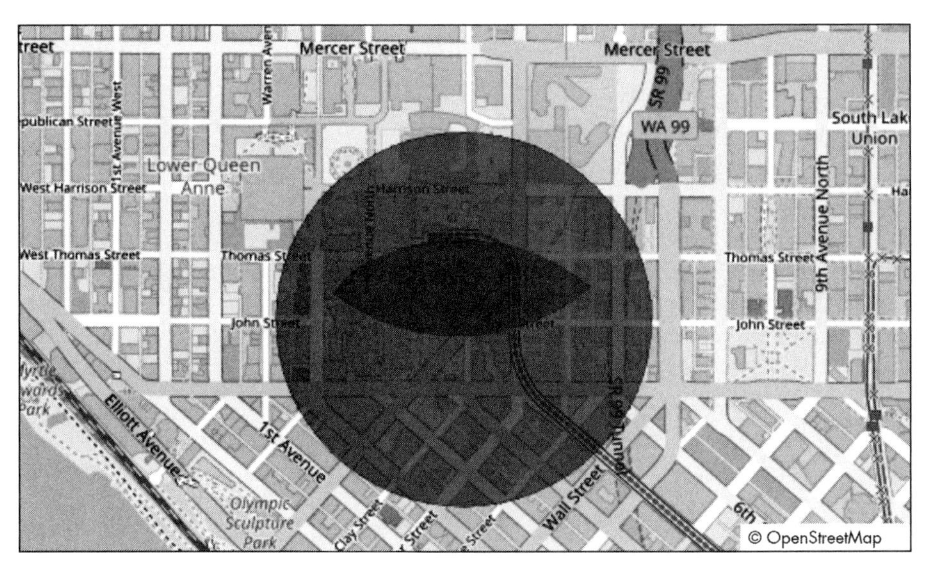

Figure 8-6: Comparing location estimates

NOTE *The code to generate the overlay in Figure 8-6 is in the 15th cell of the* OpenCell_API_Examples.ipynb *notebook.*

The light gray outer circle shows the search area provided by the OpenCellID API based on three of the four sample towers (it was actually more accurate after one of the towers was removed). The dark gray area near the middle is our search area produced using basic computational geometry. As you can see, we have reduced the overall search area considerably. Also worth mentioning is that the OpenCellID result is centered on the very edge of our search area, which means my actual location was farther from the center of the search area.

Reducing the Search Area

Using cell tower data will never be as accurate or reliable as GPS, but there are still some techniques we can apply to improve the result. By improving the signal coverage, you'll get more accurate search areas. They may actually be smaller in area, but your confidence will be higher, leading to better resource usage.

To further reduce the search area, you can take advantage of Wi-Fi networks in the area, if any are available. I wouldn't rely on them to find the initial search area, but they're a good option for shrinking a search area once you've defined one. I'm a fan of the WiGLE database for Wi-Fi searches, but the OpenCellID API also supports Wi-Fi antennas, as I mentioned previously. By combining the two APIs (and any additional APIs you may find), you'll improve your chances of finding networks with location information available. You can use the limited range of these networks to drastically reduce the search area, sometimes to a single building. The FCC has investigated using this type of Wi-Fi geolocation as one option to help emergency service dispatchers find callers who don't know their location.

You may also choose to use cell towers in your data that have the weakest signal, if you can capture the necessary information from the device you want to locate. Ideally, you'll find several towers with weak signals. These usually turn out to be the towers that are farthest apart and thus create the smallest overlapping region. This is purely heuristic, though, since a weak signal could also indicate a closer tower with more obstructions. If you can capture multiple antennas (say 12 different towers), you can try an iterative approach by testing different groups of three to four antennas at a time. You can compare the resulting search areas and determine a sort of search area heat map, where the most likely places are the ones that show up in the highest number of polygon intersections (or the overlap of the overlaps, if you prefer).

Summary

The power of geolocation in a security context can't be overstated. In a world of cell phones, evolving smart cities, and the proliferation of the Internet of Things (IoT), people are constantly saturated by network transmissions. As you've seen, an intrepid researcher, corporate overlord, or motivated hacker can take this information and turn it into a physical location. Combine that with the common business practice of helpfully naming

Wi-Fi access points after the company, and it becomes a scarily accurate tracking tool. There are a lot of ethical and legal concerns you should consider before you deploy this type of tracking system beyond a research environment.

There are several data sets that encourage users to contribute up-to-date information on the towers in their area through a process called *war driving*. Despite its antisocial name (rooted in hacker history), war driving is simply traveling around an area recording what networks are visible. Some folks have even attached recording devices to their outdoor animals so that, as the animal wanders, it also contributes to the owner's network map.[10]

It's not all gloom and doom, though. In the next project we'll look at applying the same principles of converting physical locations into geometric data to help a city plan for new emergency services. We'll revisit the topic of tessellation and discuss one of my favorite geometric algorithms, the Voronoi diagram.

9

COMPUTATIONAL GEOMETRY FOR SAFETY RESOURCE DISTRIBUTION

In enterprise security work, you'll often be asked to assist with a wide variety of infrastructure planning and deployment tasks that have more to do with providing safety than the traditional CIA (confidentiality, integrity, availability) triad we commonly think of as guiding information security. But fear not—with the mathematical tools you're collecting, you'll be able to adapt to the changing challenges and thrive under them. In this chapter we'll focus on one of the most commonly applied tools from computational geometry, and one of my personal favorite algorithms: Voronoi tessellation.

We're going to help the city of Portland, Oregon, plan the location of a new fire station using the locations of the current stations to inform our risk assessment. I picked this project because it shows how applied security concepts can scale; the same type of analysis can be applied to police stations, hospitals, burger joints, or any other public resource distributed throughout a city, state, country, or geographic region, making this one of the most flexible analysis tools in your arsenal.

This is the high-level plan: first we'll create a polygon that represents the city, and then we'll place points within the polygon representing the locations of the current fire stations. We'll split up the city into smaller polygons that represent the areas closest to a given fire station. Finally, we'll compare the areas of the smaller polygons to determine which fire station is responsible for covering the most area, and make our recommendation for the new fire station accordingly to improve response time in that area. Along the way, we'll examine Voronoi tessellations and discuss some of the limitations of our implementation. By the end of this chapter, you should have a solid understanding of how to use computational geometry for resource distribution plans. You should also feel comfortable retrieving and manipulating geospatial information using the OpenStreetMaps API, which will allow you to scale your resource planning to whatever geographic size you need.

Using Voronoi Tessellation for Resource Distribution

You already saw tessellation in action when we placed the security assets around the park in Chapter 7. In that case we divided the plane into triangles based on the vertices of the polygon, then placed a point in some of these regions representing the guard. We would expect the guard to respond to any incident that was within their zone. A *Voronoi tessellation* works in reverse, where we have a set of points (called *seeds* or *generators*) distributed throughout a plane, and we want to split up the area into regions that contain a single point. Take a look at Figure 9-1, which shows an example of a Voronoi tessellation.

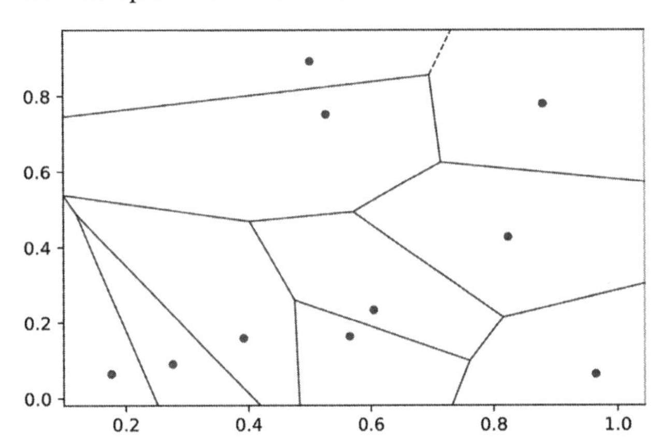

Figure 9-1: Randomly generated Voronoi tessellation

NOTE *The code to generate Figure 9-1 is in the 2nd cell of the* MapBox_Tessellation .ipynb *notebook.*

The gray circles represent 10 randomly selected (x, y) coordinates that serve as the generators for our Voronoi tessellation. Each line segment represents the place in the plane that is equidistant from two or more generators. Points along the lines are the same distance from more than one generator point, so we mark them as the border between generators. After marking all the borders between generators, we end up with a polygonal mosaic. Each polygon represents a subregion of the plane. In each polygon region there's exactly one generator point, so we can classify any arbitrary point in the plane that doesn't fall on a border line segment based on which generator point is closest.

In the park example from Chapter 7, this would be like spreading the guards through the park and then dividing up the park into areas of responsibility based on their positions. Approaching the problem this way is useful in a lot of cases, especially when examining the distribution of resources that are already in place or can't be moved easily.

NOTE *One of the most famous uses of the Voronoi tessellation came before it was formally defined and named, when British physician John Snow used the concept in 1854 to illustrate how the majority of people who died in the Broad Street cholera outbreak lived closer to the infected Broad Street pump than to any other water pump.*[1]

Formally speaking, a Voronoi tessellation divides a plane into regions of space where all the points in one region are closer to the generator for that region than any other generator. Points that are equidistant from two or more generators define the boundary between the regions. To perform the Voronoi tessellation for the fire station project, we need to define a *metric space X*, which is simply a set (in this case, of generator points in a plane defined by a 2D polygon) and a metric function that operates on the set (here, a function *d* to calculate distance between points). Our metric space—that is, the plane we'll be tessellating—is bounded by the shape of the city limits. The generator points will be the locations of the current stations that divide the city up into areas of responsibility. Both the boundary data and the generating points need to be in the same coordinate system. Rather than having to manually project the coordinates, like we did in the previous project, we'll take advantage of a purpose-built library called geovoronoi, which handles the back-and-forth projections under the hood.

The Proof of Concept: Analyzing Fire Station Coverage

The proof of concept for this project is pretty straightforward. Our goal is to produce an application that shows the current division of fire service coverage for the city of Portland, using a Voronoi tessellation to define the service areas programmatically. We want our program to find the largest service area and produce this as the recommended area to split up. To achieve this, we need to define three pieces of information. First, we need shape data representing the area we plan to analyze, in this case Portland. I've included a copy of this data in the file *portland_geodata.json* in the book's supplemental materials. You can also get the data from web services like OpenStreetMap API, which is how I retrieved it initially. The second thing we need is the location of our generator points, which, for this project, are the addresses of the various fire stations around Portland. I've included the 10 addresses I used for this analysis in the file *station_addresses_portland.csv* in the supplemental materials. Third, we need to define the function we'll use to measure distance between points when performing the Voronoi tessellation.

Once we've defined these three pieces of information, we're ready to perform the Voronoi analysis. Then it's just a matter of finding the generated region with the largest area; for this, we'll again rely on Shapely. We're going to start by discussing the distance function. Loading the area shape and generator points are interesting data retrieval tasks, but the distance function defines how the borders between regions will be calculated, so it's where the bulk of the math magic happens.

Defining the Distance Function

The metric space contains the distance function $d(p, q)$, which is used to determine the distance between points in the plane. This is how the algorithm decides which points belong to which regions. There are several choices for the distance function: Manhattan distance, Chebychev distance, sum of absolute difference, sum of squared difference, and more. Each has advantages and disadvantages, so we'll stick with the most basic and intuitive option, Euclidean distance or, colloquially, "as the crow flies."

The Euclidean distance between points p and q is the length of the line segment connecting them ($\overline{pq}$). If we treat the latitude and longitude as Cartesian coordinates, we can approach the problem as a 2D Euclidean geometry question, which we can solve using the Pythagorean theorem:

$$d(\overline{pq}) = \sqrt{\sum_{i=0}^{n} (q_i - p_i)^2}$$

Here n is the number of dimensions the problem is mapped to, or more generally the length of the vector that defines a point. So, if you were working in 10 dimensions, each point would be defined by a vector of length 10. In our case, we have two dimensions, so $n = 2$. We simply need to square the difference between the end point and the beginning point, sum the squared difference across both dimensions, and then find the square root of the result.

There is a drawback to Euclidean distance in this scenario: in reality, you can rarely move straight through a geographic area without worrying about obstructions like trees and buildings. Furthermore, vehicles like fire engines are confined to streets and subject to traffic and other conditions that dictate the path they take to their destination. I encourage you to expand on my simplification to make the results more accurate and useful in your own implementation. For now, we've defined everything we need to start diving into the question. It's time to start collecting the data required to define our geometric plane, starting with the shape of the city limits.

Determining the City Shape

To get the polygon that represents the bounds of the plane, I like to use a web tool from the OpenStreetMap team called Nominatim (*https://nominatim.org*). Its simple and free interface allows you to get several important pieces of information, like the place ID, the localized spellings of the name, and more. You can view the information on the website directly or request a JSON response to parse in your own programs, as we do here. Listing 9-1 shows how to request the information.

```
import urllib, requests, json
base = "https://nominatim.openstreetmap.org/search.php"
f = {"q": "Portland OR", "polygon_geojson":1, "format":"json"}
q = urllib.parse.urlencode(f)
resp = requests.get("%s?%s" % (base, q))
resp_data = json.loads(resp.text)[0]
```

Listing 9-1: Retrieving the JSON data for Portland, Oregon, via Nominatim

When we call the Nominatim API, the `q` parameter holds the query string we're searching; here, we set it to a string containing the city name and state abbreviation, `"Portland OR"`. The `polygon_geojson` parameter tells the API to return the geoJSON representation of the polygon indicating the boundaries of the city. This is the part we're most interested in at the moment, since it gets the shape data we need to define the city limits. The `format` parameter tells the API how we want the response data encoded. Whenever JSON is an option, it's a good choice for Python, as it handles the parsed data like a dictionary. We have to encode the parameter dictionary into a string that can be appended to the URL, using the `urllib.parse.urlencode` function. We then submit the query string as part of a GET request and parse the data using the function `json.loads` on the text value of the response. The value should come back as a list with one or more entries representing the places that match our query. In this case, there should be only one result, representing the city of Portland. Figure 9-2 shows the polygon.

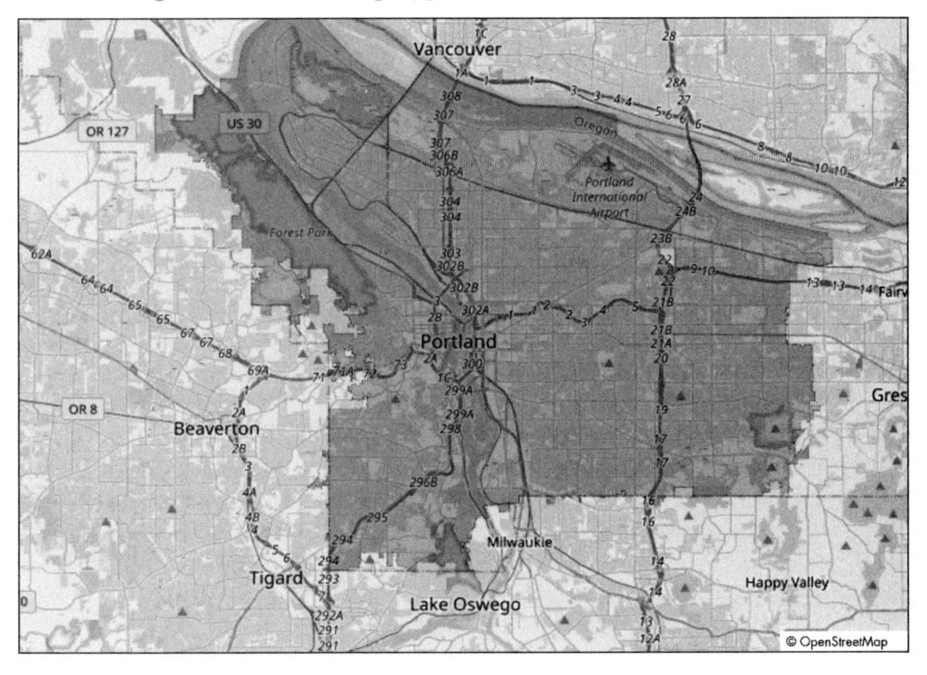

Figure 9-2: The city of Portland as a polygon

NOTE *The code to generate Figure 9-2 is in the 14th cell of the* MapBox_Tessellation .ipynb *notebook.*

The gray area represents the municipal boundaries of the city. This is the section that must be protected by the city's fire service. We need to convert the coordinates into a shape, but first let's format the data as geoJSON, as shown in Listing 9-2.

```
city_gj = {
  "type": "FeatureCollection",
  "features": [
      {
          "type": "Feature",
          "geometry": resp_data["geojson"],
          "properties": {
            "name": "City Boundary"
          }
      }
  ]
}
```

Listing 9-2: Converting the coordinates to a geoJSON feature collection

The data structure here is the top-level definition of a geoJSON object, which has the type property `FeatureCollection` to indicate that there can be a list of features nested under the appropriately named `features` key. Each feature is a nested JSON object with the type `Feature`. Each feature needs a set of coordinates that define its geometry; here we use `resp_data["geojson"]`, the geoJSON that was returned from the API in Listing 9-1. We can also add further properties to store custom information that can be used for organization or to inform our analysis. The `properties` key is followed by a nested dictionary of property and value definitions, where the dictionary's key represents the property name and the value represents the property value. Keys are restricted to string literals, but the values can be any legal JSON object, so you can get pretty creative with the property information you attach to the features.

The next step, shown in Listing 9-3, is to convert the resulting geoJSON information into `shape` objects stored together in a `GeometryCollection`.

```
from shapely.geometry import GeometryCollection, shape
city_shape = GeometryCollection(
    [shape(f["geometry"]).buffer(0) for f in city_gj["features"]]
)
city_shape = [geom for geom in city_shape][0]
```

Listing 9-3: Converting the city geometry into a collection of Shapely shapes

First we create a `GeometryCollection` by passing in a list of geometric objects making up the collection. We use a list comprehension to loop over each feature in the `city_gj` variable created in Listing 9-2. For each feature, we pass the geometry parameter to the `shape` constructor. The result is a `shape` object representing the coordinates of the geoJSON feature. Some cities are represented by more than one polygon, so this function will loop over all of the polygons that make up the city and convert each one into a `shape` object. The `city_shape` is now a `GeometryCollection` containing a single `MultiPolygon` object. We can access the `MultiPolygon` with another list comprehension. Since there's only one item, we can extract it from the list using index `0`. If you had more `MultiPolygon` objects to process, you'd want to loop over each individually. The `MultiPolygon` data in the `city_shape` variable now represents the plane we'll be tessellating using the metric space we've defined. Now it's time to get the locations for the stations to create our list of generators.

Gathering the Locations of Existing Fire Stations

As I mentioned previously, we'll be using the geographic location of the existing fire stations as the regional generators in our analysis. The file *station_addresses_portland.csv*, provided in the book's supplemental materials, contains the names and addresses of 10 fire stations within the city limits. The pandas library offers a convenient function for loading data from a CSV file into a `DataFrame`, shown in Listing 9-4.

```
import pandas as pd
stations_df = pd.read_csv("station_addresses_portland.csv", names=[
    "name", "street", "city", "state", "zip"
])
stations_df["addr"] = stations_df.apply(row_to_str, axis=1)
```

Listing 9-4: Loading the data into a `DataFrame` *the quick-and-dirty way*

Although there are libraries specifically for handling CSV data, I prefer to leverage pandas's built-in `read_csv` function, as it is flexible and comes packaged in a library I already use often. When I was preparing the data for this project, I compiled the station address list manually, using Google Maps. I didn't bother to include a header, which is why we pass in a list of columns using the `names` parameter. In an actual consultation, I'd expect the client to provide the addresses for you, so you'll likely need to adjust the file parsing to match the data format you're provided. Next we create a convenient `addr` column to hold the whole street address as a string by applying a function called `row_to_str`, which simply returns the concatenation of the `street`, `city`, `state`, and `zip` columns separated by a space character. We pass `axis=1` to tell pandas we want to apply the function to the whole row, instead of all the values in a column. We'll use the `addr` column to make it easier to search the geolocation API.

Once we've gathered the addresses into a `DataFrame`, we can again leverage the OpenStreetMaps API (via the geocoder library) to turn these into geodesic points. In Listing 9-5 we define the `locate` function to turn a single address into location info.

```
import geocoder
def locate(addr):
❶ g = geocoder.osm(addr)
    data = g.json
❷ if data is None:
        return None
❸ return {
        "address": data["address"],
        "lat": data["lat"],
        "lon": data["lng"],
        "osm_id": data["osm_id"]
    }
```

Listing 9-5: Converting an address to a geodesic point

The main work of calling the API is abstracted away for us by the `geocoder.osm` function ❶. Passing in a query string (in this case, the address) returns an object containing the API's response. The library supplies a convenient `json` parameter on the response object. If the JSON response is `None` ❷, we return `None` to indicate that the API couldn't locate anything for the input query. If a JSON object is returned, we grab a subset of the data containing the important information from the response, such as the latitude and longitude, and return it as a dictionary ❸. We also record the `osm_id`, so we can use it to shortcut future lookups or choose between multiple results for a query.

To collect the points for all the stations, we'll call the `locate` function in a loop over the addresses and store the results as a list. Since we previously created a `DataFrame` from the station data (the `stations_df` variable from Listing 9-4), we can leverage pandas's `apply` function to handle the messy work behind the scenes. We'll then convert the location information into its own `GeoDataFrame` object. Listing 9-6 shows how to handle the conversion.

```
import geopandas as gpd
locations = stations_df["addr"].apply(locate)
locations = [a for a in list(locations) if a is not None]
loc_df = pd.DataFrame(locations)
geo_df = gpd.GeoDataFrame(
    loc_df,
    geometry=gpd.points_from_xy(loc_df.lat, loc_df.lon)
)
```

Listing 9-6: Creating a `DataFrame` from the station locations

After we call apply, the locations variable contains a list of dictionary objects, one for each station, to hold the geodesic coordinates we need for our analysis. We filter out any instances where the location is None and then use this data to create a location DataFrame named loc_df. Finally, we can convert this regular pandas DataFrame into a more suitable GeoDataFrame object from the GeoPandas library. Figure 9-3 shows the results plotted on the map.

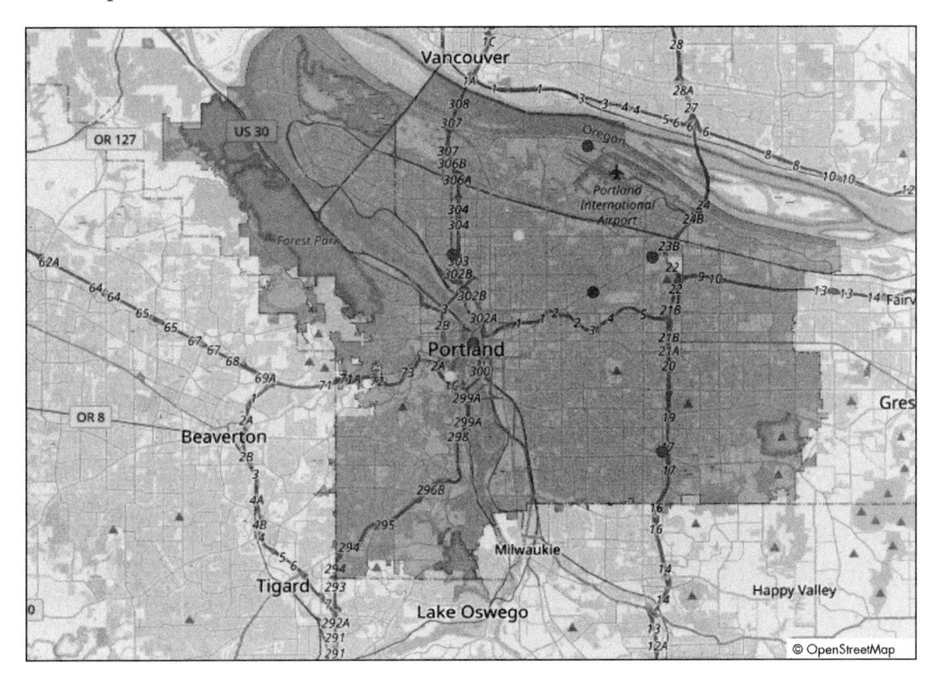

Figure 9-3: The fire station locations as points

NOTE *The code to generate Figure 9-3 is in the 15th cell of the* MapBox_Tessellation .ipynb *notebook.*

The dark circles inside the previously defined polygon show the station locations. If you count the locations on the map, only 8 of the 10 stations are present. The locate function failed to locate coordinates for two of the addresses. If this project were meant for production, we'd probably want to have multiple sources from which to retrieve the coordinates to increase the likelihood of success. You could also look them up manually or ask a client to provide the missing information. For now, we'll just drop these two from the analysis and move on.

Now that we've collected all the necessary data, we're ready to perform the actual tessellation.

Performing the Voronoi Analysis

As usual, there are several options for performing a Voronoi tessellation in Python, but the easiest by far when working with geographic information is a library named geovoronoi, which handles the coordinate projection and boundary work for us. It takes a list of coordinates and calculates the Voronoi regions using SciPy behind the scenes. At the edges of a typical Voronoi diagram, the region boundaries continue out to infinity, which isn't always the desired behavior, so the geovoronoi library allows us to take the shape of the surrounding area (in this case, the shape of a city as a polygon) to cut the Voronoi regions so that they fit into the provided shape, making the regions at the edges finite. The library also uses Shapely for managing the shape manipulation operations, making it a perfect fit for this project.

Listing 9-7 shows how to use the library along with the previously collected data to create the tessellation.

```
import numpy as np
from geovoronoi import voronoi_regions_from_coords
points = np.array([[p.y, p.x] for p in list(geo_df["geometry"])])
poly_shapes, pts, poly_to_pt = voronoi_regions_from_coords(
    points, city_shape
)
```

Listing 9-7: Converting an address to a geodesic point

The first step is to convert the points representing the fire stations into a NumPy array using a list comprehension to iterate over the `geometry` column of the previously created `geo_df` `DataFrame` object. We can then call the `voronoi_regions_from_coords` function from the geovoronoi library with the array of points (`points`) as the first argument and the boundary polygon (`city_shape`) as the second.

The result of the function is a tuple containing three useful pieces of information. The `poly_shapes` variable holds a list of Shapely `Polygon` objects representing the shape of the Voronoi regions created during the tessellation algorithm. The `pts` variable holds a set of `Point` objects representing the coordinates of the generators. These are for convenience if you haven't already created them using GeoPandas or some other method. The `poly_to_pt` variable holds a nested list for each region in `poly_shapes` that contains a list of indices into `pts`. The indices indicate the generators that belong to this Voronoi region. Usually, this is only a single point for our problem, because multiple fire stations shouldn't share the same location, but there may be cases where this isn't true and multiple generator points have the exact same location. In these situations, all of the points will be indexed in `poly_to_pt`.

Figure 9-4 shows the regions that were created along with the generators for each region.

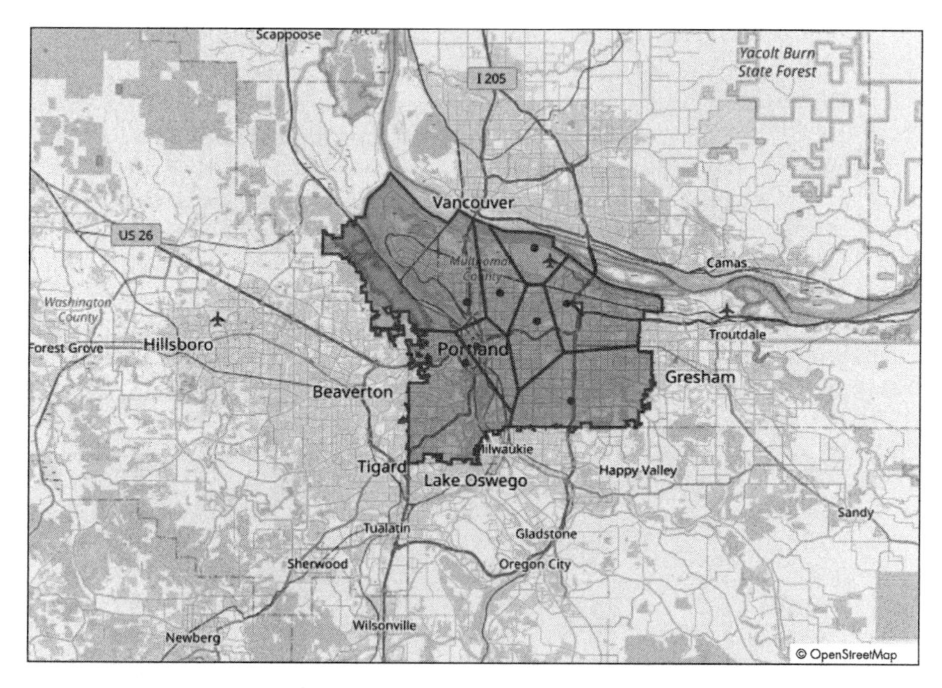

Figure 9-4: A Voronoi tessellation showing each station's area of responsibility

NOTE *The code to generate Figure 9-4 is in the 19th cell of the* MapBox_Tessellation .ipynb *notebook.*

The eight regions are divided by the black outlines. The area inside each polygon can be viewed as the naive area of responsibility (AOR) for the fire station that generated the region. It is naive because it doesn't take into account obstructions (like the water) that might make a different station the better choice to respond. Still, the information is a good starting point for you to build on. We can already see how the distribution of the stations near the center of the city leaves the edge stations with the largest AORs. Visually, the top-left AOR appears to be the largest, but we can easily verify this using the area parameter, as shown in Listing 9-8.

```
winning = 0
winner = -1
❶ for i in range(len(poly_shapes)):
    ps = poly_shapes[i]
    if ps.area == winning:
      ❷ if isinstance(winning, int):
            winner = [winner, i]
        elif isinstance(winning, list):
            winner.append(i)
  ❸ elif ps.area > winning:
        winner = i
        winning = ps.area
```

Listing 9-8: Finding the largest AOR

We begin by looping over each shape in the `poly_shapes` variable, iterating over the length of the list ❶. Looping this way allows us to keep track of the index as the function progresses. There may be cases (albeit unlikely) where more than one region has the same area. In such a case, we convert the `winner` variable into a list and keep track of the ties ❷. In the more likely scenario, the two areas aren't equal, so we check if the current area is greater than the current winning area ❸. If so, we update the `winning` amount and the `winner` index. Once all regions have been checked, the `winner` variable will contain one or more region indices, which we can use to look up the shape (or shapes) in `poly_shapes`. We can also use it to look up the station (or stations) in `poly_to_pt`.

Figure 9-5 shows the region plotted to the map along with the station location responsible for it.

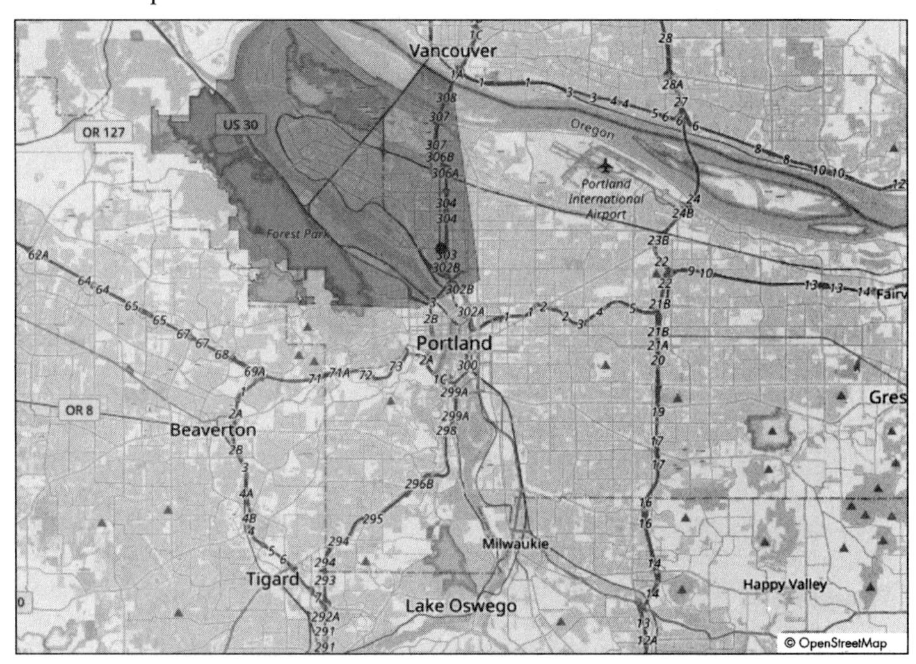

Figure 9-5: The largest region and its generator location

NOTE *The code to generate the overlay in Figure 9-5 is in the 23rd cell of the* MapBox_Tessellation.ipynb *notebook.*

Once again, the darker gray area represents the region of interest and the dark gray circle shows the location of the responding fire station. Therefore, you could reasonably argue that placing another station somewhere within that region would improve response time and resource availability in that area.

You can run the proof-of-concept code by navigating to the chapter's supplemental materials directory and running the *Emergency_service_poc.py* script like so:

```
python Emergency_service_poc.py
```

Once the code finishes loading the data and performing the Voronoi analysis, it will open a browser tab to *http://127.0.0.1*, which displays the solution using Plotly and the MapBox API.

Limitations of the Algorithm

In this chapter we've focused solely on the geographic portion of the problem, which is an excellent starting point. The narrow scope does present some limitations, though. I've already mentioned the limitations of the distance function, but there are other factors to consider if you're asked to make a recommendation like this in the real world.

For one, not all fire stations are equipped the same. Some have more trucks, some have different types of specialty equipment (like a plane for surveying large wooded areas, or a boat for doing harbor patrols), and so on. The diversity of equipment means that different stations may be better equipped to handle different problems. Putting a specially equipped station in a region where it isn't needed would be a terrible waste of resources. Another problem with the distribution of resources is that some stations are more equipped to deal with larger regions. For example, a station with a large area but 10 fire trucks at its disposal may be able to patrol its region more effectively than a station with a medium-sized region but only two trucks. You can improve upon this analysis by adding more information about the resources and specializations for the stations. You would want to ask your contacts in the city about the equipment plan for the new station and then factor that in when creating the Voronoi regions.

The final consideration is that fire stations don't see these boundaries. A fire in one region may bring responders from nearby regions as well. The fire department does a lot to try to distribute the load for any given area across two or more stations so that large fires can be controlled more quickly. The larger the fire, the more resources it will take to battle, and therefore it may be necessary to direct resources from one region to assist in another. Strictly dividing the region into AORs may lead you to suggest a station location that perfectly distributes the AORs but nevertheless doesn't allow the new station to assist the existing stations in a meaningful way.

To avoid this limitation, I suggest performing multiple analyses. For each station in the data, you can check how each other station affects its region by removing them one at a time from the list and recalculating the regions. This is similar to asking, "How would the responsibilities for station A change if station B weren't able to respond?" By overlaying the resulting regions, you'll see which stations take on the most shared responsibility because that region will have the largest overall change in area after all the other stations have been removed. A station with a lot of shared responsibility may suffer from equipment fatigue or physical exhaustion, so you may suggest that another station located somewhere closer to the overworked station could distribute the load better. Again, the station with the most shared responsibilities may have the proper equipment and personnel to handle the extra work.

All of this is meant to show you that, although Voronoi diagrams are extremely useful, they're not always the final answer. As with all analyses, the more accuracy and detail you add to your model, the more applicable the results will be to the real world. When performing any type of resource analysis, it's just as important to understand how those resources interact with their region as with each other. Doing so will allow you to make intelligent choices for your model and overcome some of the limitations of this basic framework.

Summary

In this chapter, we've covered one of the most famous computational geometry algorithms, the Voronoi tessellation. We've seen how it can be applied to real-world problems related to the distribution of resources and how it can be scaled to fit the problem. We've also discussed some of the limitations to this implementation and ways you can improve upon it yourself. My hope is that you'll take this framework and expand upon it to suit your own projects.

There's a lot of excellent research material available on the different applications of these tessellations, from security to neurology and everywhere in between. I recommend you read the research paper "Rationalizing Police Patrol Beats Using Voronoi Tessellations"[2] for another example of applying this analysis to improve emergency services. As a security analyst, you'll find plenty of opportunities to show off your resource distribution knowledge. In the final part of the book, we'll revisit tessellations to plan the distribution of security resources for the art gallery problem.

In the next chapter, we'll conclude our look into the world of computational geometry for security with one of my favorite projects of all time, facial recognition systems. Although the size and geometry are drastically different from anything we've dealt with so far, the basic ideas are the same. We'll continue to use Shapely to handle the geometry, but now we'll add some machine learning to the mix, giving us the tools needed to build the highly sophisticated analysis required to recognize facial features programmatically.

10

COMPUTATIONAL GEOMETRY FOR FACIAL RECOGNITION

Let's leave behind the world of resource distribution and look at another field where computational geometry can help you: facial recognition. *Facial recognition* is the process of examining the features of a face and determining if it matches a previously seen face and, if so, to what degree. Most babies can recognize familiar faces by three to four months old, but unfortunately, just because babies can do it doesn't mean it's easy, at least not for computers. We'll need to combine computational geometry and machine learning algorithms for our program to perform at a similar level.

We'll begin this chapter with a brief look at facial recognition. We'll then cover the main algorithm, and, as usual, put it to work in the proof of concept, which will cover loading, cleaning, and modeling data stored in a database of facial images. We'll process each image in the database to extract the most important facial features for determining which person is in an image, and then we'll use this information to build a model that can match a never-before-seen image of a face to the appropriate person.

To achieve this we'll be relying on various machine learning (ML) tools. ML algorithms seek to identify or "learn" the relationship between input data and output values using two broad categories of solutions: unsupervised learning and supervised learning.

In *supervised learning* we provide the algorithm a set of input data along with the proper output we'd like the program to learn for that input (called its *class*). In our case, the input data will be a face's geometry and the class we want to predict is the associated person's name, which makes this a *discrete classification* problem. If the class value we wanted to predict were a continuous number (such as the price of a house), this would be called a *regression* problem (you may already be familiar with linear regression from statistics class).

Unsupervised learning seeks to discover previously unknown relationships and largely deals with clustering data together based on different inputs to find interesting groupings. Because we don't know the goal going in, unsupervised learning is also sometimes called *exploratory analysis*. We won't be doing much unsupervised learning in this project, so I'll leave it to you to dive into this topic more on your own. By the time you complete this chapter, you should feel comfortable working with image data, extracting geometric facial characteristics, and training supervised learning classifiers for your own facial recognition projects.

Uses of Facial Recognition in Security

Today it isn't hard to find examples of facial recognition being used in security. Facial recognition systems have become a widespread and generally accepted part of life. Industries that are already benefiting from the technology include retail stores, casinos, cell phone manufacturers, and law enforcement agencies. It wasn't always this way, though. The core of the technology has been slowly progressing since the 1960s, when Woodrow Wilson Bledsoe invented a way for people to manually encode the geometry of a face using an electronic surface and a conductive stylus. Using the stylus, users would mark a set of standardized facial features like the bridge of the nose, eyebrows, and chin. The program would then measure the geometry between these points and shapes and create a geometric map of the input face. The data was then compared against a database of previously recorded traces to produce the name of the person the face was most closely matched to. Of course, this method was time consuming and prone to operator error. For the next 40 or so years, researchers kept improving on the algorithms by defining more standard points to measure and using image analysis techniques to automatically identify these key points in a picture of a face.

It wasn't until the turn of the century that facial recognition started to move from science fiction and research labs to reality. One highly publicized case happened in January 2001, when the city of Tampa, Florida, used a facial recognition system to record and analyze the face of every attendee during Super Bowl XXXV in the hope of spotting criminals with warrants issued for their arrest. The program is credited with identifying a few petty criminals in attendance, but is largely considered a failure due to the high cost and large number of false positives. Worse, it prompted a large backlash from privacy advocacy groups including the Electronic Frontier Foundation and the American Civil Liberties Union.[1]

The negative publicity did little to slow down the growth of the technology, though. Florida once again made the news for being one of the first states to adopt facial recognition technology as an accepted tool for police. In 2009, the Pinellas County Sheriff's Office announced a program that allows officers to tap into the photo archives of the Florida Department of Highway Safety and Motor Vehicles. Within two years an estimated 170 deputies had been outfitted with cameras that could be immediately cross-checked against the faces in the database. Since then, the availability of cheap processing power, shrinking data storage costs, and the large number of facial data sets have allowed the technology to move from government programs into the security programmer's toolbox. We'll discuss more about the privacy and ethical concerns in the next section, "Ethics of Facial Recognition Research."

Now you can build an effective facial recognition system for the cost of a decent camera (the $20 to $40 Raspberry Pi camera modules work great for this) and some cloud processing costs. I'm going to stay platform-agnostic with the concepts, but every major cloud service provider has some offering that allows you to translate the code we'll write into a distributed scalable version (we'll discuss cloud deployments more in Chapter 13).

The hardest part is collecting the database of images. For a good facial recognition data set, you need multiple pictures of the same person under different lighting conditions and with different facial poses. The features of the faces need to be distinguishable, so contrast is also important. We'll discuss more about image quality in the section "Processing Image Data" later in the chapter. For now, the takeaway is that the quality of the images in the data set will have a drastic impact on an ML algorithm's ability to distinguish between faces. Old, blurry, grainy photos with low contrast will make the process much harder, if not impossible.

We'll be using a facial recognition data set published by the computer science and engineering department at the University of Essex.[2] One section of images in the data (labeled *faces94*) is fairly stationary. Researchers had the subjects sit at a fixed distance from the camera and asked them to speak while a sequence of images was taken. The speech introduces facial expression variation, which allows the underlying classification algorithm to understand how the facial shape changes for an individual and gives it a better chance of properly classifying an input image, even if the face is in a pose the algorithm hasn't seen previously. The second part of the data set (labeled *faces95*) is more dynamic and introduces variation in scale, perspective, and lighting by asking the subject to take a step toward the camera as a set of 20 images was taken with a fixed-placement camera. The movement forward causes the head to appear larger in the later photos. It also changes the cast shadow and highlights on the face. Finally, the background for these images is a red curtain, which also introduces a degree of difficulty because the imperfect surface can make it challenging for the algorithm to detect the edges of features. Being able to properly classify faces despite all this variation will allow your program to operate more reliably in the wild, where you may not always be able to get a clean, stable image with a solid, still background.

Ethics of Facial Recognition Research

Before we move on, let's talk about the ethics of facial recognition research. There are many potential uses for, and abuses of, being able to automatically identify a person based on an image of their face. Facial recognition software can be used by first responders to identify victims after a disaster, or it can be used by dictators to identify political activists. As an analyst and developer, you'll have to be very cautious when dealing with facial recognition projects in the wild. You never really know whose hands your software may end up in.

To start, you should become familiar with all the privacy laws that might apply to your situation. International regulations like the European Union's General Data Protection Regulation consider biometric data such as facial analysis models *personally identifiable information*, which requires stricter security controls around its collection, processing, and storage. Facial recognition has even been banned from use by police forces in some US cities, like Boston, Massachusetts, due to the high error rate and potential for serious repercussions in the event of a mistaken identity. Knowing what laws and regulations apply to your project will help you navigate the other ethical questions more easily. The best way to avoid ethical and legal troubles when it comes to privacy is to gain informed consent from the people being included in the data. I strongly urge you to decline any project where you aren't able or allowed to collect informed consent.

Aside from outright abuses, there are other ethical issues that are harder to spot. Racial and ethnic biases remain a key concern of developing facial recognition models. Facial recognition algorithms often achieve mean classification accuracy over 90 percent, but researchers have shown that this error rate doesn't apply equally across all demographic groups. Several independent tests found the poorest accuracy for facial recognition was consistently for dark-skinned black females between the ages of 18 and 30. These unintentional biases are the result of technological and social choices made by the data collectors. Decisions like the type of camera lens and the locations where data was collected all have a subtle but definite impact on the overall representation of the population in the data.

I've included this facial recognition project here, despite these ethical concerns, because I think it presents an excellent learning opportunity. We'll be using publicly available data that was gathered with the consent of the individuals and the knowledge that the images would be used for research like ours, which is perfectly ethical. The data set I selected is relatively small and represents a limited number of demographics. This could lead us to make falsely optimistic performance predictions, so we wouldn't want to use it for developing any type of production system. It will serve as a good starting point, though. We can use it to illustrate the workflow and even test some parts of the algorithm. When you're ready to develop a facial recognition system for a real project, you'll have the knowledge and tools to collect a truly great data set that accurately reflects the diversity of humankind.

The Facial Recognition Algorithm

Despite the fact that we're using ML algorithms, the core of the facial recognition process remains remarkably similar to the one Bledsoe created more than 50 years ago. We'll use a set of 68 facial points to create geometric maps for a database of faces, then use those maps to train an ML algorithm to compare an input face with previously seen faces and predict the closest match. Generally speaking, facial recognition is a *computer vision* problem: it deals with teaching computers to recognize information encoded in visual data, like pictures and video. This also falls into the broader category of *multiclass classification* problems, where the class to be predicted is from a set of three or more potential classifications. When a multiclass algorithm runs, it compares the input to the data recorded for each class and determines which class the input is most likely to belong to. We're going to treat each individual person in the data set as a class we're interested in learning to predict, and the input will be an image containing the face of a previously analyzed person. The algorithm we'll use is a supervised learning algorithm; again, this means that when we train our algorithm, or model, it will have access to a list of the correct classifications, which it will use to correct previous mistakes and improve future predictions.

There's a large number of potential classes (222 unique individuals in the final analysis) and a relatively small sample size for each class (approximately 20 images per individual), making our goal even more difficult. To counter this, we'll collect a huge number of statistics for each image and let the algorithm decide which subset of these measurements allows for the best decision-making power.

Using Decision Tree Classifiers

The classification is handled by an algorithm called a *random forest classifier*, which is an expanded version of a *decision tree classifier*. There are many benefits to decision tree algorithms: they are fast to train, can produce a human-readable model, and perform well for multiclass problems like facial recognition. Let's examine a classic example as a way of illustrating how they work. Suppose we want to write a program that predicts if a person is likely to go out and play a round of golf on a given day, based on the weather. A decision tree would be a great choice for this type of problem because it will generate a list of rules we can use to examine the weather on any given day. Consider the decision tree in Figure 10-1.

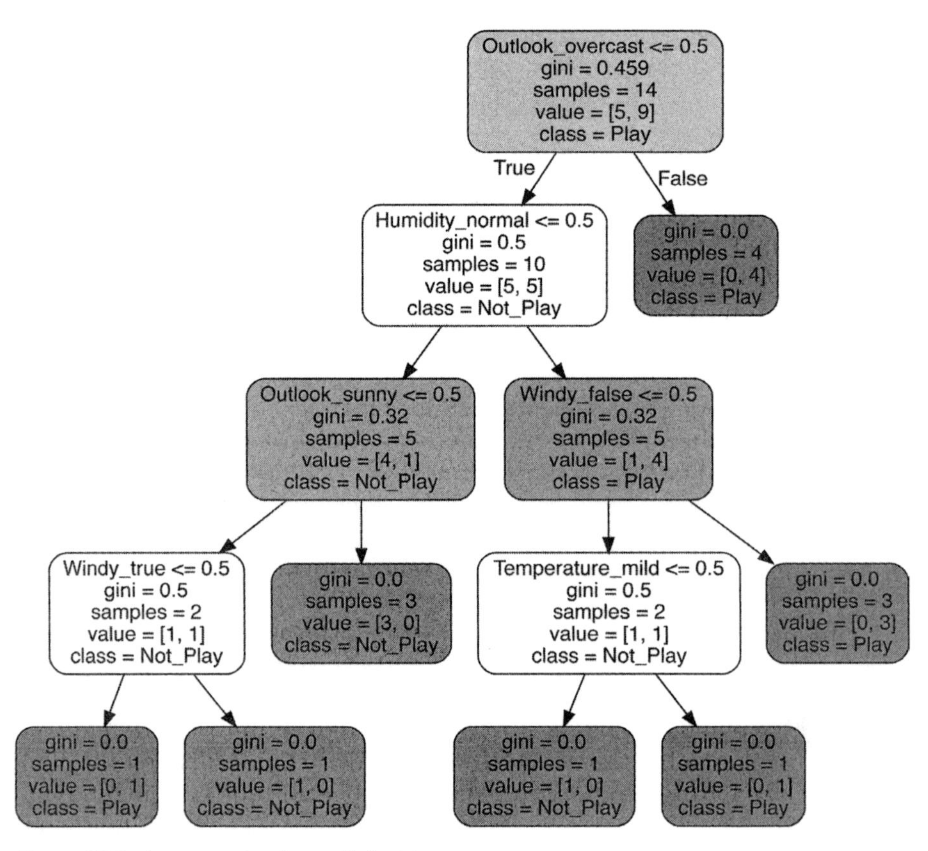

Figure 10-1: An example of a golf decision tree

NOTE *The code to generate Figure 10-1 is in the* golf_decision_tree.ipynb *notebook.*

You can see in Figure 10-1 that each branch represents a Boolean decision in the data (such as `Outlook_overcast <= 0.5`). To read the tree, you start at the topmost box (called the *root node*) and follow the proper logical branches until you reach the bottommost box (called a *leaf* or *leaves*). Some statistics about the underlying data that generated the decision are listed below the Boolean decision. Each row (called an *instance* in data science parlance) is sent through the decision tree one at a time until it ends up at one of the leaves; along the way, the algorithm records how the data influenced the growth of the tree. The simplest statistic is `samples`. This is the total number of rows that reached this point in the decision tree during creation.

In Figure 10-1 you can see that the root node received 14 samples of data. Since this is the root node, it processed every row in the data set, so this can also be interpreted as a summary of the data at the start of the algorithm. You can see the count of each class in the `values` statistic. For each possible class in the data, there's an integer representing the number of rows within that particular class. In our example there are two potential classes, `Not_Play` and `Play`. Looking at the root node again, you can see the values `[5,9]`, which means five of those samples belonged to the `Not_Play` class and nine belonged to the `Play` class.

The `gini` statistic contains the *Gini impurity* coefficient, which you can think of as a measure of the distribution of classes that reach that particular node, called the *purity* of the node. Formally, the Gini impurity coefficient can be written as

$$gini = 1 - \sum_{i=1}^{n} (p_i)^2$$

where n is the number of classes in the data and p_i is the probability of an instance being classified to the ith class. The resulting score for a node can range between 0 and 1. If all the instances in a node belong to a single class, then it is completely pure and will receive a Gini score of 0. A score of 1 means that the instance classes are randomly distributed and there's no predictability. Scores between these two extremes denote some level of class purity, with lower scores being purer (and therefore better for decision-making purposes) than higher ones. The goal is to find pure leaf nodes that contain only one class of data (a Gini score of 0). That would mean the logic that led to that leaf was capable of making a perfect decision between classes.

To see an example of this, follow the `False` branch from the root node (down and to the right in Figure 10-1). You can tell this is a leaf node because there's no Boolean expression at the top of the box and there are no branches extending from it. The `class` statistic shows the majority class for each node; when dealing with a leaf node like this, we can think of it as the probable class for data that reaches that point. With all that in mind, we can interpret this branch logically as "if the weather is overcast, predict the `Play` class" since we're dealing with Boolean values (0 or 1) and the decision criteria is `Outlook_overcast > 0.5`. The `samples` and count and the `values` statistic show that four samples reached this leaf node, all of which were of the `Play` class.

In Figure 10-1, none of the leaf nodes have more than a single class in them, making this a perfectly pure tree. Of course, more often than not, this isn't the case, and one or two stragglers from other classes (called *outliers*) may show up in a leaf where the majority is of another class. In these cases, you can try to find additional data splits that would improve the purity of each leaf, but at some point you'll have to accept the performance as "good enough" since it's unlikely you'll find a perfect split.

In the context of our facial recognition problem, we'll classify an input face by converting it to geometric information and then sending that information through the decision tree until it lands at a leaf node, which will use the majority class to predict the likely subject who matches the input face. Although the Boolean decisions the algorithm makes are more complicated than `Outlook_overcast > 0.5`, the principle remains the same.

The problem with a traditional decision tree is that it's susceptible to the data's initial conditions and configuration because it processes the samples in order. Rerunning the same decision tree algorithm on a shuffled version of the same data is likely to produce a significantly different tree each time. This means if you plan to use a decision tree, you'll need to train several versions with different mixtures of data to make sure the performance is repeatable (in this context, *performance* refers to the accuracy of predicting each class). This led researchers to design random forests. A random forest algorithm repeatedly creates individual decision trees with semi-randomized starting data (called *bagging*). To classify a new data sample, the instance is sent down each generated tree and the resulting class predictions are tallied. Finally, the majority class from all the trees' guesses is predicted as the most likely classification.

Having a large number of decision trees generated from different data mixtures will help ensure that the overall prediction is less susceptible to the starting condition of any given tree. We'll discuss the random forest more once we start building the model in the proof of concept, but before we get there, we need to cover exactly how we're going to collect the data we need. Let's turn to converting a picture of a face into a set of geometric data. In the next section, we'll discuss how to find important facial features and convert them into numeric representations.

Representing Facial Geometry

The first step in defining our facial recognition application is to figure out how to divide an image of a face into measurable shapes. I mentioned before that we'll use 68 points in an image to mark the features of the face. To save on development time and achieve our goal with less upfront coding, we'll leverage a previously trained ML model, `shape_predictor_68_face_landmarks` (*http://dlib.net*)[3], which identifies the points of interest in a frontal view of a human face. Figure 10-2 shows the 68 points laid approximately where they would fall on a face.

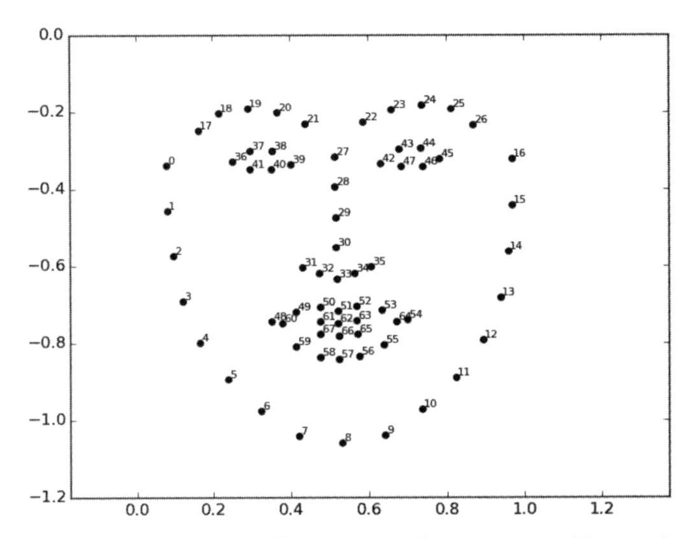

Figure 10-2: The points of interest on a face, generated by an algorithm (image source: https://i.stack.imgur.com/OBgDf.png*)*

The map of geometric features includes the jawline (points 0–16), left and right eyebrows (points 17–21 and 22–26, respectively), nose (points 27–30 for the bridge and 31–35 for the base), left and right eyes (points 36–41 and 42–47, respectively), and the mouth (points 48–59 for the exterior of the lips and 60–67 for the interior). When the algorithm receives an image of a face, it adjusts the locations of each point to try to match the positioning of the input face. We'll use the adjusted locations of these points to create Shapely shapes representing the different features. We'll then compute some geometric statistics about the face, such as the distance between the eyes, the length of the nose, and so on, to create a statistical representation of the face. Listing 10-1 shows how to load the model.

```
import dlib
detector = dlib.get_frontal_face_detector()
predictor = dlib.shape_predictor(
    "facial_model/shape_predictor_68_face_landmarks.dat"
)
```

Listing 10-1: Loading the facial landmark detector

The model is part of the dlib library, which wraps several C++ functions in Python goodness so you can leverage the speed of C++ for scientific computation and the friendly syntax of Python for everything else. The `get_front_face_detector` function returns a previously trained model for detecting faces in images based on a method known as *histogram of oriented gradients (HOG)*. The detector counts occurrences of gradient orientation in localized portions of an image, meaning it examines only a small box of pixels at a time; this is similar to the way a human might examine a picture with a magnifying glass to focus on detailed areas (except there's no distortion of the pixels in this case). The output of the `get_front_face_detector` function is a list of (`index, rectangle_coordinates`) tuples, one for each detected face. We store this information in a variable named `detector`, which we'll use to help the predictor focus in on the faces. The actual facial feature location is handled by a shape predictor that takes in an image region containing some object (in this case, a face) and outputs a set of point locations that define the pose of the object. To load the predictor, we tell dlib the path to the model we're interested in, which, in this case, is *facial_model/shape_predictor_68_face_landmarks.dat.*

NOTE *There are many other models available from researchers you should look into as well, including ones that can detect faces from a three-quarter view or even detect other types of objects such as cats or dogs.*

With the face detector and landmark detector components defined, we're ready to start processing images. It's very important to the accuracy and reliability of your facial recognition system (and any other predictive algorithms, for that matter) that you process the test images exactly the same as the training images; otherwise, the difference in processing may corrupt the results in unpredictable ways. The next section will describe a modular bit of code we can use in the proof of concept to handle both the training data creation and the test image processing independent of the modeling functions.

Processing Image Data

There are many possible ways to process image data so that predictive algorithms can build a model based on their information; they all share the same goal, however, of converting the data into a normalized format. The way you intend to predict faces has a lot of influence on what processing steps, if any, you should take before converting the picture into a feature set. For example, you may choose to preserve color information by creating a predictor for each of the three color channels (red, green, and blue) individually, in which case you wouldn't want to convert the image to grayscale, as we do here. Other operations are fairly common regardless of the final processing plan. Operations such as cropping the image to the facial region or resizing the image help ensure that all samples are consistently scaled and the features are somewhat normalized in the end. Listing 10-2 shows the function for processing images in the *.jpeg* file format.

```
import cv2
import imutils
def process_jpg(file_path):
    img = cv2.imread(file_path)
    image = imutils.resize(img, width=300)
    return cv2.cvtColor(image, cv2.COLOR_BGR2GRAY)
```

Listing 10-2: A function for processing a single .jpeg *image*

We start by defining the `process_jpg` function. The only parameter we need is the path to the *.jpeg* image, stored in `file_path`. We use the cv2 (short for *computer vision 2*) library's `imread` function to read the file into a data array representing the pixel values for each color channel. Then we resize the image data using the `imutils.resize` function. We scale the images so they have a width of 300 pixels using the `width` parameter; the height of the image will be calculated based on this new width to avoid distorting the features. Finally, we convert the resized image data to grayscale using the `cv2.cvtColor` function and return the result. (I chose to convert the image data to grayscale since color information won't help inform our geometric analysis.)

NOTE *The accompanying Jupyter notebook for this chapter also has a similar function to process* .gif *images, which you may come across in some facial recognition data sets. The imutils and cv2 libraries work well together for a large number of image processing and computer vision tasks, so I encourage you to dive into both libraries more and see what other preprocessing steps you might apply.*

Figure 10-3 shows an example result.

Figure 10-3: Processed facial images ready for analysis

NOTE *The code to generate Figure 10-3 is in the 6th cell of the* Facial_Recognition_notebook1_Processing.ipynb *notebook.*

You can see that we get back grayscale images scaled to 300 pixels wide by 450 pixels tall. Note that the facial features aren't distorted; this is because of the scaling method we used. We'll use these two images to exemplify the rest of the process. Both images have decent contrast and the facial features are not obstructed (by things like sunglasses, hats, or heavy makeup), making them good candidates. Both also have some areas that will ultimately prove more difficult for the algorithm, as you'll see.

Resizing the images and adjusting the color to grayscale before we continue processing them will allow us to get consistent, repeatable samples to work from, regardless of slight differences in picture quality, scale, and lighting. These steps are just the beginning. You should consider other image transformations you could apply (such as increasing the brightness or playing with the contrast values) to give the algorithm the best chance of success when analyzing the image. In the next section, we'll take our processed images and begin the work of actually locating and analyzing the facial features.

Locating Facial Landmarks

Now that we've defined our image processing step as a function, we can call it at the beginning of our facial landmark code to ensure that we're working on a grayscale version of the resized image. Our next task, shown in Listing 10-3, is to write the function that will locate the facial landmarks we'll use to define the rest of the facial structure.

```
from imutils import face_utils
def locate_landmarks(image_file):
 ❶ gray = process_jpg(image_file)
 ❷ clone = gray.copy()
 ❸ rects = detector(clone, 1)
   feature_coordinates = {}
```

```
❹ for (i, rect) in enumerate(rects):
    ❺ shape = predictor(clone, rect)
    ❻ shape = shape_to_np(shape) # See PoC code
    ❼ for (part_name, (i, j)) in face_utils.FACIAL_LANDMARKS_IDXS.items():
          if len(rects) >= 2:
              feature_coordinates[part_name] = []
          for x, y in shape[i:j]:
            ❽ feature_coordinates[part_name].append((x, y))
      face_points = []
      for n in feature_coordinates.keys():
        ❾ face_points += feature_coordinates[n]
    ❿ feature_coordinates[part_name] = face_points
  return feature_coordinates
```

Listing 10-3: Locating feature landmarks in an image

The `locate_landmarks` function takes in a filepath as its only argument. Then, we start by calling the `process_jpg` function from Listing 10-2 to get the processed image from the file argument ❶. I prefer to work on a copy of the image ❷ to avoid accidentally overwriting the source with any changes. Once we've copied the image, we locate the rectangular region for each face in the input using the `detector` we created in Listing 10-1 ❸. The result is a list of tuples containing the index of the face and the coordinates corresponding to the rectangular area that contains it. For the sample data there will only ever be one rectangle in the list, but you could extend the function to handle the case of multiple faces for a future project.

Next, we loop over the list of rectangles ❹ and send each to the `predictor` we also set up in Listing 10-1 ❺. The result is the list of point coordinates in the order shown in Figure 10-1. We use the function `shape_to_np` ❻, which is a simple helper function to convert the shape's (*x*, *y*) coordinates into a NumPy array.

```
def shape_to_np(shape, dtype="int"):
    coords = np.zeros((68, 2), dtype=dtype)
    for i in range(0, 68):
        coords[i] = (shape.part(i).x, shape.part(i).y)
    return coords
```

Here we create a NumPy array of zeros, named `coords`, to hold the 68 coordinate pairs. Next, we loop over all of the indices in the `shape.part` list. For each part we create a tuple from the `x` and `y` attributes and assign the tuple to the `coords` array at the same index. Once we've collected all the coordinate pairs, we return the resulting array of tuples.

Back in Listing 10-3, to make things easier going forward, we've built a dictionary that is keyed off the feature name and contains the list of points defining the feature. The `face_utils.FACIAL_LANDMARKS_IDXS.items` function ❼ returns a list of tuples that conveniently provides the feature name and a nested tuple that defines the start and end index in `shape` that correspond to the feature. We loop over each of these definitions and create a corresponding entry in the `feature_coordinates` dictionary ❽. If there's more than one face in the picture, the index number of the face will be appended to the feature name to keep them separated.

Next, we create a list of all the points in the face data and append it to a list ❾; this will be used to calculate the convex hull of the face in the proof of concept. Formally speaking, a *convex hull* is the smallest convex polygon that encloses a set of points such that each point in the set lies within the polygon or on its perimeter. Remember from Chapter 7 that a convex polygon is one where all its interior angles are less than 180 degrees. You can think of convex hulls as the result you'd get if you could stretch a rubber band around the outside of all the points in a shape.

Finally, we add the resulting list to the `feature_coordinates` dictionary ❿.

Figure 10-4 shows the results of running the algorithm against our two test images.

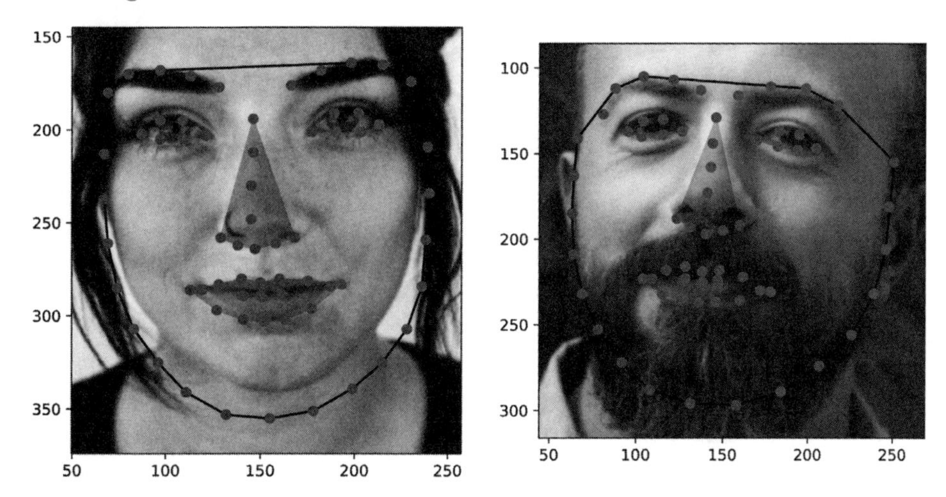

Figure 10-4: The results of landmark detection

NOTE *The code to generate Figure 10-4 is in the 9th cell of the* Facial_Recognition_notebook1_Processing.ipynb *notebook.*

The gray dots are located where the algorithm believes each feature of the face is placed. The black outline around the entire facial region represents the resulting convex hull. As you can see, the program did a decent job. For the most part, the dots correctly hit the landmarks we're looking for. Where the program fell short is finding the jawline on the woman in the left image, because the contrast between her dark hair and the background is more defined than the difference between her jawline and the background. The measurements of her face are going to be skewed, which, unless they're skewed the exact same way every time, will result in bad training data.

The algorithm did a much better job finding the man's jawline, even though it's hidden behind his beard. But the algorithm has failed to find the edges of his nose properly. It's a fairly small difference from the actual position, however, and though it will impact a few measurements, the overall face shape remains in proportion and thus would generate usable training data. After a few months of doing these tests and examining the results, you'll be able to look at an image and closely estimate how well the landmark detector will perform given the processing steps you define.

After calling the `locate_landmarks` function, you should get back the dictionary keyed by the feature name, which allows you to refer to the collection of points using a human-friendly name. For example, to create polygons representing the left and right eyes, you can use this code:

```
from shapely import Polygon
leye = Polygon(feature_coordinates["left_eye"])
reye = Polygon(feature_coordinates["right_eye"])
```

You can then use Shapely to measure the minimum distance between the shapes:

```
dist = reye.distance(leye)
```

Or you could measure the difference in area between the two eyes:

```
diff = abs(reye.area - leye.area)
```

There are many other potential characteristics you could use, but we'll dive into those more when we start to build the training data set in the next section. Now that we've defined the functions we need, we can start building the proof of concept for our facial recognition system.

The Proof of Concept: Developing a Facial Recognition System

The proof of concept for this project is separated into two parts. The first part builds the training data from the set of facial images. Here, we'll prepare each image for processing and define the statistics to collect from it. This is where computational geometry is going to help us. The image processing steps may take several minutes (or longer with much larger data sets), so it makes sense to compute this on its own and store the results to a file for processing later. This spares us from having to run expensive calculations over and over. It also makes adding new images to the data set easier. Rather than rebuilding the entire data set to retrain the model, we only need to process the new images before retraining. Be warned: trying to process all of these images in memory isn't going to work. We'll need to process the files one at a time to keep memory usage manageable. You'll see how later in the chapter.

The second portion of the proof of concept defines the ML algorithm and trains it on the previously computed statistical data. We'll test the algorithm multiple times using a cross-validation process called *leave one out (LOO)*. We'll run the validation once for each class in the data, during which the LOO algorithm selects an image from that class to hold out from the training data (hence the name). The model is then trained on the rest of the data. After training the model, we'll give it the selected image to classify, then total the results for each class to estimate overall performance. The major benefit of the LOO method of validation is that it provides the training algorithm with the most information because only a single instance is removed before training. Since we have a limited number of images for each face, we need to give the training algorithm the best chance possible to succeed.

Facial Statistics

The first part of the proof of concept is in the *facial_recognition_poc_1.py* file. It covers the code you've seen up to now, but we'll expand on it to create the final training data set. There are a number of ways you could approach collecting statistical information about the facial structure. My first attempts involved tessellating the face in different ways and measuring the predictive power of different sections. Figure 10-5 shows the method that scored the best, applied to the two example faces.

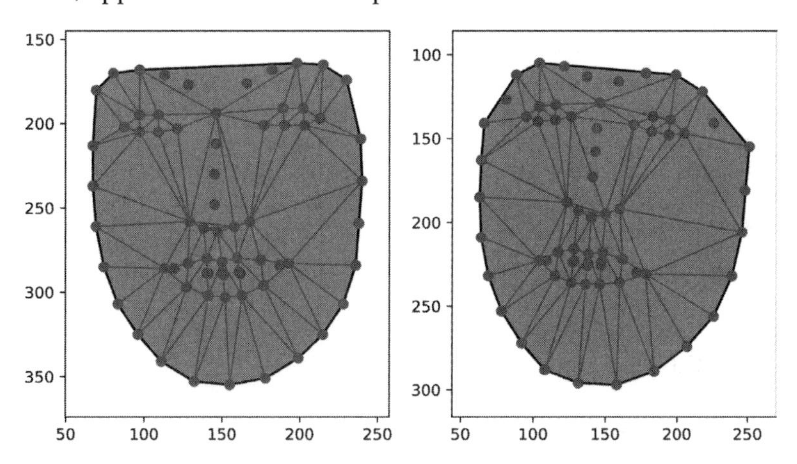

Figure 10-5: A result from automated facial tessellation

NOTE *The code to generate Figure 10-5 is in the 10th cell of the* Facial_Recognition_notebook1_Processing.ipynb *notebook.*

The tessellation treated the nose, eyes, and inner mouth portions as holes in the main facial polygon. The primary problem with this method turned out to be too much noise from similar triangles that didn't contribute to the structural knowledge at all (for example, all of the triangles making up the chin); this created a large number of similar-valued variables across all the faces. Another problem with this method is that tessellation doesn't include all the shapes of interest. For example, creating a triangle from the outer points of both eyes and the bottom of the chin would indicate the tilt of the entire head. Why am I bothering to tell you about this if it didn't work? Because it's important to realize that trial and error is necessary. Thinking about why a method failed can be more informative than considering why it works!

During my research I found a paper from the Facial Identification Scientific Working Group (FISWG) that does an excellent job describing the standard facial statistics.[4] Ultimately, I changed tactics from automatic tessellation to explicitly defining 62 measures taken mostly from the reference material. To help with defining the measurements in the code, you can create some variables that represent key points by name. For example:

```
nose_btm = feature_coordinates["nose"][6]
bridge_top = feature_coordinates["nose"][0]
upper_lip_ctr = feature_coordinates["mouth"][3]
lower_lip_ctr = feature_coordinates["mouth"][9]
chin_ctr = feature_coordinates["jaw"][8]
r_temple = feature_coordinates["jaw"][0]
l_temple = feature_coordinates["jaw"][16]
```

You can use these points to create the measurements in a way that allows you to return to the code months later and still understand which variables relate to which points of the face. Listing 10-4 shows a sample of the metrics from the proof of concept.

```
  from shapely import LineString
  face_dict = {}
❶ face_dict["tri_area"] = Polygon([r_temple, chin_ctr, l_temple]).area
❷ face_dict["face_vert"] = LineString((chin_ctr, bridge_top)).length
  face_dict["bow"] = LineString((upper_lip_ctr, nose_btm)).length
❸ face_dict["bow_ratio"] = face_dict["face_vert"]/face_dict["bow"]
```

Listing 10-4: Defining the geometric statistics with Shapely

The three major types of statistics we collect are areas, distances, and ratios. Area metrics convert a set of facial points into a polygon object and then record the `area` property of the shape ❶. Distance metrics create a `LineString` object from two or more points and then record the `length` property ❷. Ratios are derived metrics that compare the values of two previously created metrics. For example, here we've compared the length of the line between the upper lip and bottom of the nose (colloquially called the Cupid's bow) to the total vertical height of the face (measured from the chin to the top of the nose) ❸. Ratios should compare only statistics of like types. The ratio of one area to another area or the length of one distance to another makes sense, but the ratio of a distance to an area doesn't make much sense in this context.

On top of the 62 explicit metrics, I've included the x-coordinate and y-coordinate as separate features in the data for a total of 214 data points per image. We'll let the model-building algorithm determine which of these features are most informative in the "Feature Engineering" section of this chapter, but for now, defining more statistics means a higher chance of finding some meaningful ones.

Memory Management

As mentioned previously, trying to process all of these images in memory isn't feasible. Images contain a large amount of information, and trying to open a large number at the same time for processing will quickly fill your memory buffers. Instead, the proof of concept opens a single image at a time and computes all the statistics. It saves the data and immediately closes the image to move on to the next. Listing 10-5 shows the structure of the loop.

```
❶ image_paths = get_image_files("faces95", image_paths)
  face_collection = []
  for image_file in image_paths:
    ❷ feature_coordinates = locate_landmarks(image_file)
    ❸ if len(feature_coordinates.keys()) < 8:
          continue
      --snip--
    ❹ face_series = pd.Series(face_dict)
      face_collection.append(face_series)
❺ faces_df = pd.DataFrame(face_collection)
❻ faces_df.to_csv("facial_geometry.csv")
```

Listing 10-5: Looping over image files for processing

The get_image_files function is another helper function in the proof of concept that recursively walks a directory structure collecting filenames that don't end with *.txt* ❶. Once we've done that, we loop over each filepath in the resulting list and pass it to the locate_landmarks function we defined in Listing 10-3 ❷. Some images in the data set aren't clear enough for the landmark detector to find all the features. In these cases, the feature dictionary won't have the proper number of keys, and we can skip any further processing ❸. The snipped section is where we'll add the code to create all of the facial statistics using the method shown in Listing 10-4.

Once all the data points are created, we convert the dictionary into a pandas Series object ❹ and append it to the collection of faces. After all the images are processed, we create a DataFrame from the list of Series objects ❺. Finally, we save the results to a *.csv* file for later use ❻. Running the script will conclude the first portion of the proof of concept by creating a data set derived from the geometric statistics. We'll use this data to train the classifier we develop in the next two sections.

Data Loading

At this point we've created our geometric data set and saved it to a *.csv*. The second stage of the proof of concept is located in the file *face_recognition_poc_2.py* and picks up with loading this previously created data into pandas. Listing 10-6 shows how to load the *facial_geometry.csv* and prepare the data for the association calculations.

```
  import pandas as pd
❶ faces_df = pd.read_csv("facial_geometry.csv")
❷ faces_df.drop(["Unnamed: 0", "file"], inplace=True, axis=1)
❸ faces_df["category"] = faces_df["name"].astype("category")
❹ cat_columns = faces_df.select_dtypes(["category"]).columns
❺ faces_df[cat_columns] = faces_df[cat_columns].apply(lambda x: x.cat.codes)
❻ name_map = faces_df[["name", "category"]].set_index(["category"])
```

Listing 10-6: Preparing the facial_geometry.csv *data for training*

We start by calling the pandas `read_csv` function to get the data from the previous step ❶. The creation of the file results in an `Unnamed` index row, which we drop to save memory ❷. The next step is to define the `category` variable. The `name` field contains a randomly assigned fake name to make the data realistic while still preserving the data subjects' privacy. This is the column we're interested in building a model to predict, but pandas treats it as a text string by default, so we convert it to a categorical column using the `astype` function with the type `category` ❸. We collect the name of all the categorical columns using the `select_dtypes` function. The result is a list of column names in `faces_df` that have the type `category` ❹. Currently the `category` column should be the only result in the list, but it still makes it easier to reference all categorical variables this way should you choose to add more categorical information in the future.

Because pandas automatically assigns a numeric index to each category derived from a categorical column, we overwrite the content of the `category` column with that numeric index using the `apply` function ❺. For convenience, we create a lookup table so we can convert category IDs to names or vice versa by copying the `name` and `category` columns from `faces_df` and saving them into another `DataFrame` object called `name_map` ❻.

That wraps up our data loading code. At this point, we've loaded the previously created *facial_geometry.csv* file and converted the categories, which are the subjects' names, into a format pandas can understand. The next step is to set aside a *true holdout set*, or the instances from the data that are never used during the feature engineering, training, and performance estimation phases. I chose three instances as the size of the holdout set so that the model training portion had plenty of data left to learn the model from. One major problem with ML occurs when researchers accidentally give the algorithm direct or indirect access to the answers for the test data; this biases the model, so it performs excellently on the test set but will likely fail horribly in practice.

As an example, suppose all subjects have 20 images except the three subjects that are held back for a true holdout set, which have only 19 images remaining in the data. If the algorithm had access to the count of images for each subject, it could narrow down the list to only those three subjects, even though this information isn't likely to translate well in a production system where subjects will have different numbers of photos to work from. So, to make sure we haven't tainted our result, the algorithm will hold back the three samples from different classes (one picture from three different people) to test the final model against. Using a true holdout set is equivalent to testing with three never-before-seen pictures, which is as close to testing on production requirements as you're going to get.

In Listing 10-7 we create the three-instance holdout set.

```
from random import choice
real_test = {}
while len(real_test) < 3:
❶ name = choice(list(faces_df["name"].unique()))
    if name not in real_test.keys():
    ❷ group = faces_df[faces_df["name"] == name]
```

```
        ❸ real_test[name] = choice(group.index.to_list())
❹ index_list = [r[1] for r in real_test.items()]
❺ real_X = faces_df.iloc[index_list]
❻ faces_df.drop(index_list, inplace=True)
```

Listing 10-7: Randomly selecting a true holdout data set

We use the choice function to randomly choose a subject name from a list of unique names ❶. It would be better to do this by ID for real data, since the probability of two people in a corporate data set having the same name is fairly high. Luckily, we don't have to worry about this in the sample data because the names were generated using faker (a library to randomly generate data that looks authentic), so the probability is much lower. If the selected subject is already in the holdout set, we continue to choose names randomly until we find one who isn't.

Next, we gather all the instances for the randomly selected subject ❷. We use the choice function once again to select a random instance index from the group of instances ❸. The result is a dictionary keyed off the subject name with a value that indicates the randomly selected instance's index. Once the indices have been collected, we use a list comprehension to collect all the indices from the dictionary ❹. We copy the actual instance data from the faces_df object into a separate DataFrame ❺. Finally, we remove the instances from faces_df so they won't be used in the association matrix calculations ❻.

Feature Engineering

Now that the image processing is complete, it's time to move on to the actual model training code. For this portion of the proof of concept, we're going to apply feature engineering and ML to our previously generated facial data. Our goal is to produce a predictive model capable of identifying a subject based on a previously unprocessed image containing a face it has previously analyzed the geometry for (using the holdout set we created in Listing 10-7). To achieve our goal, we need to cut the excess noise from our data set so our algorithm can focus on the really informative measurements. That's where feature engineering comes in.

One of the most important steps in any ML project, feature engineering involves mathematically analyzing the relationship between the different variables in the data and the class value we're interested in predicting to determine which ones add the most useful information. Your ability to predict anything useful is directly tied to the quality and quantity of the data available. These days, a lack of data is hardly a problem. Quite the opposite: we usually have so much data about a topic that figuring out what's really important to the outcome is nearly impossible for a human to do by hand. The important relationships can get drowned in the noise of useless data. To counter this problem, researchers use one or more feature engineering algorithms, which score the features in the data based on their contribution to some value we want to predict (in this case, the subject name associated with the face). We're going to apply three steps to progressively whittle the features down to only those we're confident are contributing to the model's accuracy.

Association Matrix

One popular method of scoring features is an *association matrix*, which determines which features in a large list of features are most correlated to one another. The result of running an association algorithm is an $n \times n$ matrix where n is the number of features in the data. Each cell contains the correlation score between the two features defined by the column and row. Figure 10-6 shows a portion of the association matrix for the facial data.

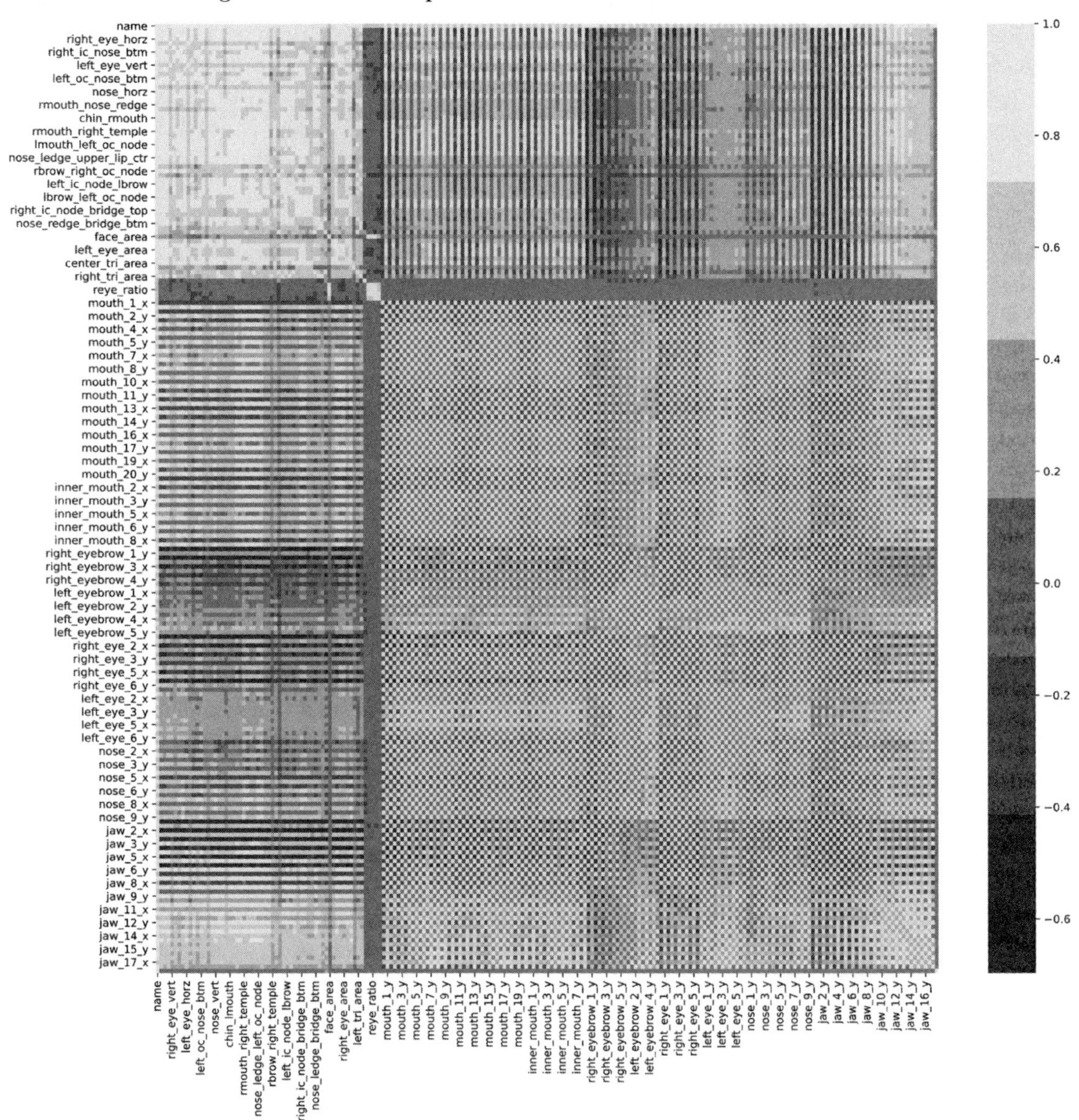

Figure 10-6: A feature association matrix

NOTE *The code to generate Figure 10-6 is in the 15th cell of the* Facial_Recognition_notebook2_Modeling.ipynb *notebook.*

The correlation between a variable and itself will always be 1.0, so you can ignore those instances. What we're most interested in are those features that have a high correlation with the feature we want to predict. Note that the correlation is taken as an absolute value to treat negative and positive correlation as equally important. Taking highly correlated variables together offers the best chance to properly predict the value of interest.

The limitation to normal correlation approaches is that you must be correlating some continuous (real) numeric values. In this case, we want to measure the correlation of continuous variables with a discrete categorical variable (a subject's name), so standard correlation measures won't work. Instead the matrix was calculated using another correlation score known as Theil's U, which handles categorical data, with code borrowed from a blog post by Shaked Zychlinski (*https://towardsdatascience.com/the-search-for-categorical-correlation-a1cf7f1888c9*). All of the functions are located in the *nominal.py* file with the author's original descriptions of how they operate, so I'll focus on how we integrate the `association` function into our facial recognition system.

Listing 10-8 shows how we can calculate the association matrix.

```
import nominal
assoc_matrix = associations(
    faces_df,
    nominal_columns=cat_columns,
    theil_u=True,
    return_results=True
)
```

Listing 10-8: Calculating the association matrix of the feature set

The `association` function from the *nominal.py* file takes in a `DataFrame` object to perform the calculation on; here, that's `faces_df`. You need to pass in a list of nominal columns. This is where the `cat_columns` variable defined in Listing 10-6 comes in handy; we won't need to edit the code, even if we add more categorical information. To use Theil's U as the calculation, we must set the `theil_u` parameter to `True`. By default, the `association` function just displays the results to the screen, but we want to use the data to programmatically select which features to use in the model, so we set the `return_results` parameter to `True` to also return the result as a matrix.

Now that we have an association score for each feature, we can collect the top predictors (those with high association scores with the `category` column) into a list so we can compare results with the next two feature engineering steps. An iterative approach may be to start with the 10 top-performing features and see if you can train a functional model. Continue increasing the number of features by 5 to 10 until you find the lowest number that produces a reliable model. Another approach (and the one I prefer) is to set a predictive threshold for the features you want to keep as follows:

```
assoc_matrix = assoc_matrix[abs(assoc_matrix["name"]) > 0.95]
key_features = [k for k in assoc_matrix["name"].index]
```

Picking a point to cull features is as much an art as a science. After some trial and error, I found that I could keep features that scored higher than 0.95 and still get good performance from the final model. The `key_features` variable contains a list of the 19 column names that have an association score greater than 0.95 with regard to the `name` column; this includes the `name` and `category` columns, which have a score of 1.0, as I mentioned earlier.

Still, the association matrix is just one indicator of predictive power. To be really sure we're picking the best set of features, we'll run another feature selection algorithm and compare the best performers from both to see which features are in both lists. Those features will have a very good chance to improve our prediction's accuracy.

Mutual Information Classification

If one measure of association is good, then two should be great, right? In this case we can apply a second method of ranking features to gain even more insight into which features will be most helpful. The *mutual information (MI)* score between two variables is a non-negative value that measures the dependency between the features. It is equal to 0 only when two random variables are completely independent. Higher values indicate higher dependency.[5] Listing 10-9 shows how to calculate the MI using scikit-learn.

```
from sklearn.feature_selection import mutual_info_classif
contributing = mutual_info_classif(
    faces_df.drop(["name", "category"], axis=1),
    faces_df["category"],
    discrete_features="auto",
    n_neighbors=7
)
```

Listing 10-9: Calculating the MI contribution for each feature

The first argument is the feature matrix we want to calculate the MI score for. To avoid tainting the data, we drop the `name` and `category` features from the set with an inline call to the `drop` function, which doesn't remove the columns from the actual data, just the temporary feature list passed into the algorithm. The next argument is the list of instance categories that will be used internally to train a classifier; we pass `faces_df["category"]` as it contains the classes of the data we're interested in finding the MI scores in relation to. You can set the `discrete_features` parameter to a list of feature labels to treat explicitly as discrete values, or you can let the algorithm attempt to detect the discrete features by setting it to `auto`, as we've done here. The results are calculated using a nearest-neighbor classifier, so once again a bit of trial and error can be necessary to find the right number of neighbors. An iterative method can help you find a good setting for this parameter. After a few test runs, I settled on seven neighbors. The result of the call to `mutual_info_classif` is a list of values in the same order as the columns in the `faces_df` data.

The Jupyter notebook has code showing how you can use scikit-learn to automate the process of iterating over model parameters using model selection to find the best configuration.

As before, we can collect the top performers by selecting all the features with an MI score greater than 1. We'll compare this to the list of key features generated from the association matrix to create an even smaller list of features that are highly informative when it comes to predicting the value in the category column. Listing 10-10 shows how.

```
results = zip(
    faces_df.drop(["name", "category"], axis=1).columns,
    contributing
)
mi_scores = [f for f,v in results if v >= 1]
reduced_features = [k for k in mi_scores if k in key_features]
```

Listing 10-10: Finding the overlap between best feature lists

The zip function combines the column names from the data with the results in the contributing variable we defined in Listing 10-9 to create a list of tuples with the structure (*column name, MI score*). We then use a list comprehension to filter the results into a list of column names that had a score greater than or equal to 1.0. Finally, we compare the column names in the key_features list to the columns in the mi_scores list. Any column that's in both lists makes it to the reduced_features list, which represents those features that have scored well on both association and mutual information with respect to the categorical variable we want to predict. At this point there should be only nine columns left in the running, so we could stop here if we wanted. We've reduced the data set from more than 200 features down to just 9. In practice, you could probably get reliable performance modeling off of these, but I like to push things a bit—let's see how extreme we can go. We'll do one more feature engineering pass to see if we can concentrate the predictive power even more.

Correlation Ratio

In statistics, the *correlation ratio* is a measure of the relationship between the statistical distribution within individual categories (in this case, geometric descriptions of photos for the same person) and the distribution across the whole population or sample (geometric descriptions of all photos in the data set). The measure is defined as the ratio of two standard deviations representing each variation, or the ratio of a feature's variation within a class compared to its variance over the whole set. An ideal feature would have low variance within a single category but high variance between categories.

Intuitively, this is the same as saying we want features that are consistent for one person but differ between people. One example would be the distance between the outside points of the eyes. For a single person, we'd expect this measurement to be fairly consistent, but we'd expect measurements between two different people to produce a larger difference.

Listing 10-11 shows how to gather the correlation ratio for the `faces_df` data.

```
etas = {}
for feat in faces_df.columns.to_list():
    if feat not in ["category", "name"]:
      ❶ etas[feat] = correlation_ratio(faces_df[feat], faces_df["category"])
❷ sorted_rank = sorted(etas.items(), key=lambda kv: kv[1])
reduced_key_features = []
for f in sorted_rank[-21:]:
  ❸ if f[0] in reduced_features:
        reduced_key_features.append(f[0])
```

Listing 10-11: Calculating the correlation ratio for the features

The `correlation_ratio` function ❶ also resides in the *nominal.py* file. It takes in a `Series` object representing the feature we want to score, along with another `Series` representing the categorical feature. The value is in the range of real values between 0 and 1, where 0 means a category cannot be determined by the feature's measurement, and 1 means a category can be determined with absolute certainty. We assign the resulting value to a dictionary keyed off the column name. Once all the features have been scored, we sort the dictionary using the `sorted` function. The result is a list of tuples with the structure (*`column name, value`*) sorted from worst performance to best ❷. We loop over the last 21 entries and compare each column name to the `reduced_features` list. If a column is in both lists, it goes into the final feature set called `reduced_key_features` ❸. The result is a list of the four most predictive features from the data:

```
['outer_eyes', 'nose_area', 'face_horz', 'center_tri_area']
```

Each of these features was in the list of top performers for all three feature selection methods, which is a strong indication of its ability to classify data. In the next step, we'll use only these four features to train our models. In a production system, you could use this information to reduce the number of statistics you collect during the first phase. Clearly most of the measurements I defined in the first phase weren't necessary to distinguish faces in the data, but in the beginning you rarely know what will be useful, so collecting a large number of data points and letting the algorithms do the work can reveal unexpected relationships. It's very important to note these features are specific to this data set. You can't do feature selection on one image set and expect those features to translate perfectly to every other image data set.

Model Training

Now the data is finally ready for modeling. We're going to create the reduced data set using the four features selected during feature engineering. We'll then establish a null hypothesis by scoring a simple classifier on the data. Finally, we'll build the real thing using the random forest classifier to perform the final classifications for the three images held out from the beginning.

Splitting the Data

The first step is to split the data into training and testing sets. Listing 10-12 defines the two data sets as well as the object to handle the test data generation.

```
from sklearn.model_selection import LeaveOneOut
X = faces_df[reduced_key_features]
y = faces_df["category"]
loo = LeaveOneOut()
splits = list(loo.split(X))
```

Listing 10-12: Creating the training and testing splits

Traditionally, the variable `X` is used to denote the test data (which doesn't contain the classification information). Here we take only the subset of the features defined in `reduced_key_features` from the `faces_df` data to define `X`. We use the variable `y` to hold the corresponding class information from the `category` column. Finally, we use the `split` method from the `LeaveOneOut` class to create a set of *n* copies of the data in `X` (where *n* is the number of distinct classes). The result is a list of tuples with the form (*`training indexes, test indexes`*) called `splits`. In the LOO validation scheme, each split has a different image removed from the data set to test on, so the test indices will always contain a single instance ID and the training indices will have the rest. As I mentioned before, the LOO validation method gives the algorithm the most training information. It's also closer to how the system would be used in production where an image of a subject's face would be compared to a facial database.

Establishing a Baseline

To establish a baseline score, we begin by modeling one or more *dummy classifiers*, which use very simple prediction methods (such as random guessing or guessing the majority class) to establish a worst-case performance score. Listing 10-13 shows the API for working with scikit-learn classifiers.

```
from sklearn.dummy import DummyClassifier
from sklearn.model_selection import cross_val_score
❶ dc = DummyClassifier(strategy="uniform")
scores = []
hits = 0
misses = 0
for train_index, test_index in splits:
  ❷ X_train, y_train, = X.iloc[train_index], y.iloc[train_index]
  ❸ X_test, y_test, = X.iloc[test_index], y.iloc[test_index]
  ❹ cvs = cross_val_score(dc, X_train, y_train, cv=4)
  ❺ score = sum(cvs) / 4 # Default cv value
    scores.append(score)
  ❻ dc.fit(X_train, y_train)
```

```
    ❼ y_pred = dc.predict(X_test)
    ❽ if y_test.values[0] == y_pred:
          hits += 1
      else:
          misses += 1
❾ print((sum(scores) / len(scores))*100)
❿ print((hits / (hits+misses))*100)
```

Listing 10-13: Training a baseline dummy classifier

We use the `DummyClassifier` class from scikit-learn (which has the same API as the actual classifiers) to define the baseline model. Passing the `uniform` argument ❶ creates a model that will randomly guess the class from the set of possible classes. We loop over the splits defined previously to create the training and testing instances from the corresponding index lists. The `X_train` and `y_train` variables ❷ are used to train the model (if the model is one that requires training), while the `X_test` and `y_test` variables ❸ will be used to score the resulting model.

We use the `cross_val_score` function ❹ to get an a priori performance estimate for the split. The function takes a classifier object and the two parts of the training data set. The `cv` parameter sets the number of folds that will be used to validate the model. The default is five folds, but the data includes a class that has only four images, so if we don't set this to `4`, Python will print a bunch of warnings. The function performs the four-fold cross-validation using the classifier object passed in. To get the average score across all four folds, we take the sum of the scores and divide it by the number of folds ❺. We save the average score into a list so we can analyze it after the loop completes.

Next, we fit the `DummyClassifier` object with the training data ❻. *Fitting* a classifier is the proper terminology for training the model on the data. Nothing happens internally when we do this for our dummy classifier, but it's the proper workflow when we go to use a more sophisticated classifier in the next step, so it's best to stick to convention.

Finally, we use the fitted classifier to predict the outcome from the `y_test` data set ❼, which is the one image held out by the LOO algorithm. The result of the `predict` function is the class the algorithm thinks the data belongs to. We compare the predicted class to the actual class ❽ and increment the hit or missed count accordingly.

Once the loop completes, we calculate the average cross-validation score by summing the `scores` list and dividing by the length of the list ❾. The result is the average of average scores for cross-validation, which is a decent indicator of real-world performance. In my tests the `DummyClassifier` scored about 0.4 percent accuracy. To validate the performance estimate, we also calculate the ratio of hits to misses ❿. During my testing, the actual score was just under 0.6 percent (26 hits out of 4,457 chances). This result serves as the null hypothesis for our conclusion going forward (if you aren't familiar with hypothesis testing, check out this article: *https://www.statisticshowto.com/probability-and-statistics/hypothesis-testing*). If our actual classifier can do better than 0.6 percent correct classification, then the model is having some impact on the correctness beyond random coincidence.

Implementing the Random Forest

Now it's time to implement the random forest classifier. Since the API is shared between all scikit-learn classifiers, the code remains largely unchanged from the dummy classifier code in Listing 10-13. Listing 10-14 shows the changes.

```
from sklearn.ensemble import RandomForestClassifier
from random import randint
❶ rfc = RandomForestClassifier(
    n_estimators=100, min_samples_split=5, min_samples_leaf=3
)
❷ for i in range(50):
    split_i = randint(0, len(splits))
  ❸ while split_i in chose:
        split_i = randint(0, len(splits))
    chose.append(split_i)
--snip--
```

Listing 10-14: A decision tree algorithm definition

Rather than defining a dummy classifier, we use the `RandomForestClassifier` class from scikit-learn's `ensemble` module ❶. *Ensemble* classifiers use multiple classifiers internally and aggregate the predictions into a single prediction. In this case, each internal classifier is a random tree trained off a random sampling of the input data—hence the random *forest*.

The `n_estimators` parameter defines the number of internal classifiers to train. The `min_samples_split` parameter defines the minimum number of instances used to train the internal classifiers. The `min_samples_leaf` parameter tells the random trees the minimum number of samples to consider a valid leaf node. Setting this parameter to higher values will start to automatically prune less useful logic branches from the resulting decision trees. If you look back at the decision tree in Figure 10-1, you can see that the leaf nodes at the very bottom of the tree have only one sample each. Since the data set was small to begin with, that's fine, but if the number of samples at a leaf node is low when there's an abundance of data, it most likely means that logic branch doesn't add as much information as other leaves with more samples covered. You can once again use manual or automated parameter tuning to find an optimal configuration here.

The second change is to use the `randint` function to randomly select splits ❷ rather than iterating over all of them in order. I chose 50 random splits for no special reason; I encourage you to find a number that works better. We use a `while` loop to ensure the split hasn't already been used to train a random forest ❸.

From this point on, the code is identical to the previous dummy classification code (you can compare cells 23 and 25 in the *Facial_Recognition_notebook2_Modeling.ipynb* notebook to see this). Just be sure to rename the `dc` object references to `rfc`. And with that, the code should be ready to run. The result of my test gave a five-fold cross-fold estimated performance of 76.5 percent and a test set performance of 72 percent (36 hits out of 50 chances). It may not seem like 72 percent is all that great, but considering the fact that we're predicting the correct outcome using only four geometric features, it's pretty impressive!

Testing the Holdout Images

For the final validation of the model, we're going to give it the three true holdout images and see if it can predict the correct subject. Remember that these images haven't been used by any portion of the code up to now, so they're completely new to the model. Given the previous result of 72 percent accuracy, it would be fair to guess the result should be two or three correct out of the three possible. Listing 10-15 shows how we run the trained classifier on the holdout images.

```
real_y = real_X["category"]
test_X = real_X[reduced_key_features]
rfc = RandomForestClassifier(
    n_estimators=100, min_samples_split=5, min_samples_leaf=3, random_state=42
)
rfc.fit(X, y)
y_pred = rfc.predict(test_X)
print(list(zip(y_pred, real_y)))
```

Listing 10-15: Testing with the true holdout data

The `category` column from the `real_X` data set defines the proper classes we want predicted. We define the `test_X` data by taking the `reduced_key_features` subset from the `real_X` data set. Then we create the `RandomForestClassifier` object exactly as before. When we fit the model, we use the entire data set not in the true holdout set. Then we call the `predict` function on the `test_X` data. The result is a list of the class indices that should match the three indices in the `real_y` list. To compare the two easily, we can use the `zip` function to combine the two lists and print it out.

Here's the result from my test while writing this:

```
[(25, 25), (122, 122), (174, 174)]
```

A perfect score! It's important to realize that, due to the stochastic nature of the algorithm, your results will change between runs. The dummy classifier in my tests ranged from 0.2 percent to 0.6 percent, while the `RandomForestClassifier` regularly scores above 72 percent. If you get an odd result, such as zero correct classifications from the holdout set, try rerunning the code. With an expected precision of 72 percent, there's still a 28 percent chance one holdout image will be misclassified, a 7.8 percent chance that two will be misclassified (0.28^2 = 0.078), and a 2.2 percent chance of the algorithm misclassifying all three holdout images (0.28^3 = 0.022).

At this point we've proven our concept is viable. You can refine the process to improve the accuracy and reliability of the algorithm, but we've proven that we could in fact use computational geometry to produce a functional facial recognition system. Clearly, the result is no fluke, given the dreadful performance of the baseline classifier, so we've achieved our goal of properly predicting three subjects from previously unseen images.

In addition to the accuracy, you should also consider adding a measure of confidence in the prediction. What we've done by predicting a single class is called a *hard classification*. The problem with it is that you'll always get back some prediction, regardless of whether there's any reason to believe it's accurate. You could instead opt for a *soft classifier*, which predicts the likelihood that a test instance belongs to any given class in the data. After fitting the classifier, you can use the scikit-learn `predict_proba` function instead of the standard `predict` to get back the list of probability-like scores. By examining these scores, you can get a sense of how "sure" the algorithm is in a prediction. You can calibrate the random forest classifier to get better probability scores and set a threshold confidence level to accept or reject a classification. You can see an example of using probability predictions in the *Facial_Recognition_notebook2_Modeling.ipynb* notebook in the section labeled "Soft Prediction."

The final step in our proof of concept is to save our work for future use. In the next section, we'll cover saving and reloading the trained model in a way that's suitable for deployment in modern production environments.

Model Persistence

It wouldn't be practical if you had to retrain the model every time you wanted to classify an image of a face. Training a model like this from any realistic facial database will take hours on a single machine. Luckily, we can store the results of the trained model and then load that saved state into one or more processing applications without having to train them on the data directly.

The scikit-learn documentation on model persistence (*https://scikit-learn.org/stable/model_persistence.html*) recommends using a library called joblib to handle storing the data in a *pickled* format. As you might know, pickle is Python's most popular data serialization and storage library and is capable of storing complex Python objects to a file on disk. The joblib library includes two functions, `dump` and `load`, which are convenience wrappers around the pickle library. The following code uses the `joblib.dump` function to save the trained model to a file named *trained_facial_model.pkl*:

```
import joblib
joblib.dump(rfc, "trained_facial_model.pkl")
```

Now we can load the previously trained model into another program by simply calling the `joblib.load` function:

```
import joblib
loaded_model = joblib.load("trained_facial_model.pkl")
```

From this point, you can treat the `loaded_model` object the same way you would treat the previously trained `rfc` model. If you run the `type` function on the `loaded_model` object, you'll see `<class 'sklearn.ensemble._forest.RandomForestClassifier'>`.

One of the major benefits to saving a trained model as a pickled object is that we separate the training of the model from the use of the model.

When we need to incorporate new data into our model (such as images of a new person's face), we can rerun the model training code without interrupting any running analysis. From the point the training process completes, all future runs of the code that loads the model can use the updated model, allowing for seamless updates.

Finally, a note on security: it's well known that unpickling a specially crafted malicious object can result in arbitrary code execution. Since the `joblib.load` function just wraps the `pickle.load` function, the same is true for it as well. To reduce your risk, you should never unpickle or load an object from an untrusted source. When you develop an application that loads pickled models in production, you need to make sure you add some type of data integrity check *before* you unpickle the model.

Summary

In this chapter we've explored all the steps necessary to develop a functional facial recognition system using our knowledge of computational geometry and a healthy dose of machine learning principles. With some tweaks and further testing, you can definitely improve the system's performance. By combining the geometric information used here with other nongeometric analysis—such as color palette histograms, wavelet transformations, or Eigenfaces—you can get the performance well above 95 percent. Some researchers have reported accuracy as high as 99.96 percent (0.04 percent error rate, as of 2020).[6] This was under optimal conditions on a test set developed by the US National Institute of Standards and Technology (NIST) for scoring vendors of facial recognition systems.

Improving the accuracy and confidence thresholds is very important when you plan to apply these facial recognition systems in a security context. Misclassifying a face that isn't in the data as one that is—in other words, a false positive—could lead to accidental authentication of something very sensitive. Another potential security risk researchers have proven under test conditions is the use of photos or specially designed masks to bypass facial recognition systems.[7] Clearly there's still a lot of open area for research and improvement in this field. To continue developing the system, you can combine all of the elements into a processing pipeline using a platform like TensorFlow or Spark to distribute the computational load across many computers and begin to fix (or attack) some of these problems yourself. You can also translate the principles into other image classification domains, such as fingerprint analysis or software failure analysis.

This ends our adventure through computational geometry. I hope the last three projects have shown you how flexible a tool it can be. Oftentimes, translating all or part of a problem into a geometric representation can help you understand the essence better. There's a vast amount of theory left for you to explore on your own. With the foundations you've picked up here, the rest will be much easier to understand and apply in meaningful security applications.

We'll revisit some computational geometry in the next part of the book, "The Art Gallery Problem," where we'll use it to help analyze the shapes of rooms and the locations of resources, much in the same way we applied Voronoi tessellation to analyze the distribution of Portland's fire stations. Let's turn there now!

PART III

THE ART GALLERY PROBLEM

11

DISTRIBUTING SECURITY RESOURCES TO GUARD A SPACE

The remainder of the book will focus on a single, extremely practical application. Known as the art gallery problem, this classic application has plenty of research for us to leverage and deals with efficiently distributing security resources to guard a space. Efficiency is an imperative for today's security teams: there are always more assets to protect than there are resources to protect them. In its grandest form, the art gallery problem ties together the two main disciplines we've been studying: graph theory and computational geometry. It's fitting, then, that this part of the book will also represent the most complete Python application yet, going beyond a conceptual design and into the realm of a full-fledged software project. We'll cover the design, development, and delivery options for a modern Python project, including graphics, distributed computation, and licensing your application to users.

The goal for this part of the book is to develop a *minimum viable product (MVP)*, which can be considered a step up from a proof of concept. As you've seen, a proof of concept proves that an idea is worth pursuing and defines the framework for future development. It's usually constrained to only the functions necessary to get an idea off the ground—no bells or whistles included. An MVP design, on the other hand, is concerned with the smallest number of features you could produce to bring an idea to market and "be competitive." Typically this means adding features like a GUI as well as user-friendly elements, such as save and restore functions.

There's no exact set of features that make an application viable, because they're ultimately dictated by the expectations of the users in that particular market. For example, in the security market, users have come to expect features like single sign-on (SSO), data encryption, push notifications, and so on. Does this mean you need to develop all of these features before you can bring a security tool to the market? Absolutely not! Think minimally and ask yourself, "What features do all the applications similar to mine share?" When you're developing a product for users, it's tempting to try to anticipate all their needs up front, but this mentality is impractical and expensive: you'll often end up solving problems that no one would have encountered (edge-case code) or developing features that only confuse new users. If you instead release a product that contains a minimal set of clearly labeled features, you can get feedback on the actual problems users see and conveniences they're missing. Iterative improvement plans allow you to focus development time on features that will actually be useful.

Now let's jump into the art gallery problem, which asks, "What is the minimum number of guards that need to be placed in a gallery (represented by an *n*-vertex simple polygon) such that all points of the interior are visible?" This is a resource planning problem similar to the fire station placement problem from Chapter 9. A good plan for the placement of security personnel, checkpoints, and monitoring devices can reduce the number of incidents a security team will need to respond to from the start. It can also improve the response time when an incident does occur, thus reducing the overall risk. Unfortunately, there are often differing levels of understanding among human planners on a security team, which can lead to poorly planned (or poorly implemented) security controls. That's why I'm always searching for ways to automate portions of my team's planning.

It was during one of these searches that I discovered the art gallery problem, which addressed the very problem I was researching: the efficient deployment of security resources for buildings with what we'll call "untraditional" layouts. As you'll see, not all building designs lend themselves equally well to being guarded, so before we get into the details of the problem, we'll cover the use cases for the application we plan to develop. Then we'll be ready to start developing the core of the application logic. We'll cover the existing research and show the theory in its simplest form. We'll then move on to defining the two data representations we're going to use in solving the problem and discuss the data structures. Finally, we'll go beyond the base model to allow for more realistic deployments by adding advanced concepts like field of view and budget constraints.

Determining the Minimum Number of Guards

We'll use the original problem statement as our first use case: a user wants to know the minimum number of guards to protect an unconventional floor plan. We want all the guards together to be able to observe the whole gallery (all points of the interior, or the walls, in the original problem statement). For this use case, our application will need functions that can encode floor plans in a data format the computer can analyze as well as an algorithm that can do the actual guard placement. We'll go through that in the rest of this chapter, once we've covered a few more use cases.

The next use case deals with informing the architectural design of secure facilities and can be summarized like so: a user wants to analyze the security coverage and layout of a building. Prior to building a secure facility, the CAD drafts of a few potential floor plans are run through a *building information modeling (BIM)* program. The hypothetical building designs are rated on the difficulty to secure, access to emergency exits, accessibility features (like ramps and elevators), and more. For this use case, the application will need to define the effective coverage for different types of security equipment, including human guards' cameras and other electronic sensors.

Of course, this analysis can also inform an attacker of flaws in a security layout. Every heist movie has a scene where the protagonists lay out the blueprint for their target and start to mark where the impossibly large number of security controls are, until they spot a flaw. Sensor blind spots are a favorite trope for these films, but the truth is that blind spots are a real-world consideration. By observing the guards and sensors in use, you can make a fairly accurate coverage map. Finding blind spots is often a simple matter of searching for the model's technical specifications document. Information such as field of view, often given in degrees, as well as effective range, given in feet or meters, tells you not only where the device can detect but also where it can't! To support this use case, the application we're developing will create a visual layout of how the gallery is divided among the security resources. A user will be able to inspect the layout to see any gaps in the coverage. We'll also develop a solver that will suggest additional vertices where guards can be placed to achieve customizable coverage goals, and we'll add the concepts of distance, field of view, and effective range so we can differentiate between a guard, camera, motion sensor, and so forth.

Many buildings, particularly galleries and museums, are multistory buildings, so we can assume a professional using our software would want it to work on all the floor plans. Because the project needs to contain multiple assets, such as the shape data and the floor plan itself, it's also safe to assume a user will want a way to back up their work or resume working across multiple sessions. We'll wrap these all together in the final use case: a user wants to plan security for multiple floors of the same building across multiple work sessions. The program we develop will do this by creating a set of custom objects that can be serialized, encoded, and stored in a compressed file between sessions. We're going to treat each floor of a multilevel building as separate 2D floor plans that can be grouped into a single multifloor project. Each floor will contain an image representing the background used to trace the gallery shape as well as all of the geometric information added by the user.

Studying blueprints and CAD designs of buildings is an excellent habit to form, especially for physical penetration testers. Simply by knowing the layout of a facility and walking confidently, I've passed myself off as belonging in a building. After all, who would know that "utility closet #2" is on the basement level, west end, if not someone who had been there before? Pair that knowledge of the layout with a clipboard, toolbox, or other official-looking props, and it's often just as good as a building pass.

Now that we have our use cases, we can start diving into the nuts and bolts. Let's start by reviewing the original problem and some research performed in the past. We'll then cover the solution steps in depth and discuss additional constraints you can add to customize the results to specific needs.

Art Gallery Problem Theory

The first well-known theorem on the art gallery problem was written in 1973 by computer scientist and professor Václav Chvátal. This question was posed to him by a former University of Washington mathematics professor, Victor Klee:

> Given the floor plan of a weirdly shaped art gallery having *n* straight sides, how many guards will we need to post, in the worst case, so that every bit of wall is visible to a guard?

The *Chvátal AGP theorem* gives an upper bound that states, "At most $n / 3$ guards is always sufficient, and sometimes required, to cover a polygon with *n* vertices."[1] Chvátal assumed in his proof that guards would be placed on the vertices, but *Chvátal's upper bound* remains true even if the restriction of guards at corners is loosened to "guards anywhere within the interior of the polygon." The 3 constant comes from decomposing the shape of the gallery into triangles, based on the reasoning that you would only ever need one guard per triangular section.

Chvátal's work was later simplified by mathematics professor Steve Fisk, who reduced the problem to a three-color problem, defining it as follows: "Under what conditions can the regions of a planar map be colored in three colors so that no two regions with a common boundary have the same color?" The three-color problem can be represented easily as a graph, where each vertex of the gallery's shape constitutes a node and each edge marks a shared wall segment between two vertices. You can then treat the problem as a *vertex-coloring problem*, where two nodes of the same color can't be directly connected by an edge. The graph version of the coloring problem is a popular method to analyze connectivity, so NetworkX includes a function to solve it for us, known as *greedy coloring*, that we'll leverage in our solution. Figure 11-1 shows the simplest case of the greedy coloring algorithm.

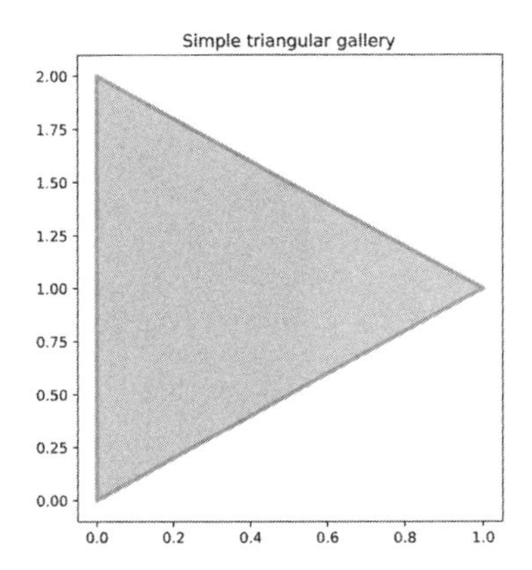

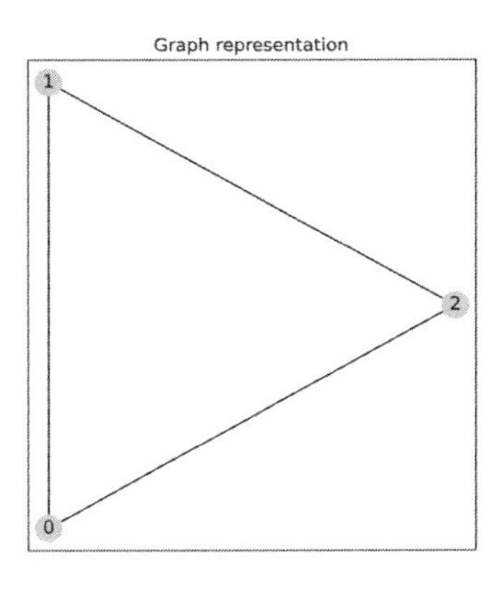

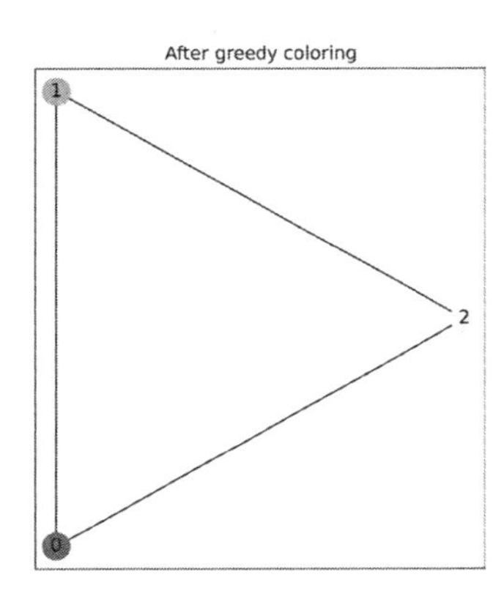

Figure 11-1: Solving the simplest art gallery problem

NOTE *The code to generate Figure 11-1 is in the 4th cell of the* AGP_basics.ipynb *notebook.*

The leftmost image in Figure 11-1 shows a triangle polygon. The upper bound of $n / 3 = 3 / 3 = 1$ means we should need only one guard to observe the entire interior. The middle image shows the result of converting the triangle into a graph representation. Finally, the rightmost image shows the result of the greedy coloring algorithm. As expected, each node gets colored a different color, meaning placing one guard at any one of those points would enable them to observe all the walls. This is the core of the process. To expand the algorithm, we just need to solve a series of interconnected triangles, as you'll see in the next section on the geometric and graph representations.

In a 2008 thesis paper, Mikael Pålsson and Joachim Ståhl examine the three-color algorithm and propose a set of alternate "rectangular" algorithms (ones that operate on orthogonal polygons only) specifically designed with camera placement in mind.[2] Pålsson and Ståhl were able to address several practical pitfalls of Chvátal's theorem. First, by restricting themselves to orthogonal polygons, they reduce the upper bound to $n / 4$. Second, their camera placement version of the problem addresses concerns like limited field of view, effective range, and obstacles, which apply equally well to human guards as well as cameras. These additional goals make the placement selection more realistic than the standard formulation. We'll discuss field of view and effective range more in a bit. We'll also add the ability to weight areas to prioritize the required coverage. Other constraints, such as making sure that each camera is visible to another camera (common in high-security areas), won't be covered but are definitely worth your time to research.

Because it uses orthogonal polygons, Pålsson and Ståhl's approach is less practical for general use in planning security layouts. Anywhere there's a diagonal or curved wall will require multiple small rectangles to approximate the shape. One of my favorite examples of unconventional architecture is the Guggenheim Museum in New York, shown in Figure 11-2.

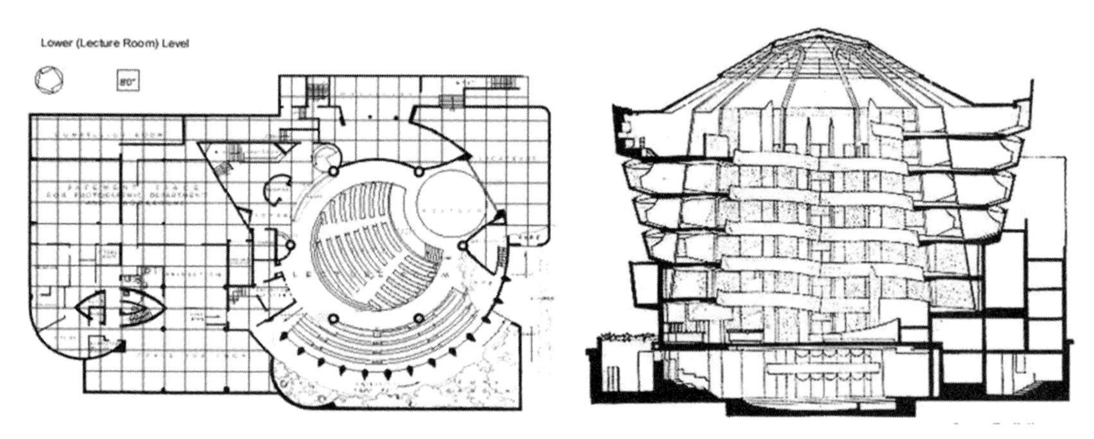

Figure 11-2: The Guggenheim, designed by Frank Lloyd Wright (© 2023 Frank Lloyd Wright Foundation. All Rights Reserved. Licensed by Artists Rights Society.)

On the left you can see a top-down view of the lower level, the lecture auditorium. The grid laid over the floor plan represents 8 square feet. As you can see, near the rounded walls and near the diagonal walls, the squares are cut off in places. These are the areas where a purely orthogonal approach would struggle. On the right you can see a cross section of the museum showing the circular structure that continues throughout the majority of the building. It makes for beautiful architecture, but it also makes the museum difficult to map with orthogonal polygons.

To counter the shortcomings of the orthogonal approach, we're going to allow users to draw one or more complex polygons directly on top of an image of the building's floor plan. We'll treat each individual polygon as a distinct area that needs to be guarded, similar to the original problem definition. Each polygon will be subdivided using a modified triangular tessellation called a constrained Delaunay triangulation. We'll then convert the tessellated geometry into an unweighted graph, which we can solve using the greedy coloring algorithm from NetworkX. This process allows users to input unconventional floor plans like that of the Guggenheim and solve them for a wide variety of potential scenarios, while still respecting the real-world limitations of the resources involved.

Geometric and Graph Representations of the Gallery

Now let's consider how we'd represent the gallery space as a `Polygon` object, just as we represented the park in Chapter 7. Figure 11-3 shows the example gallery.

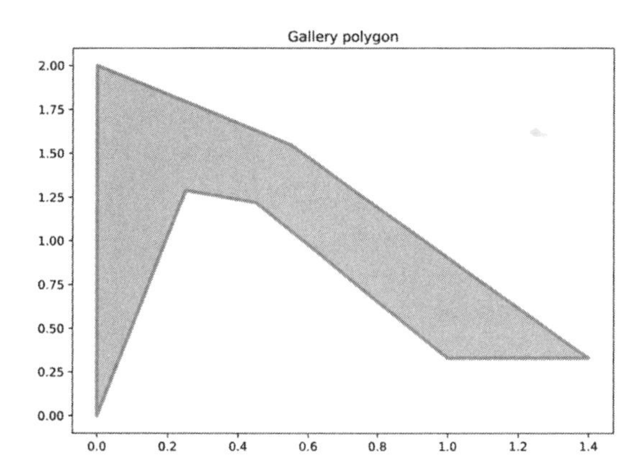

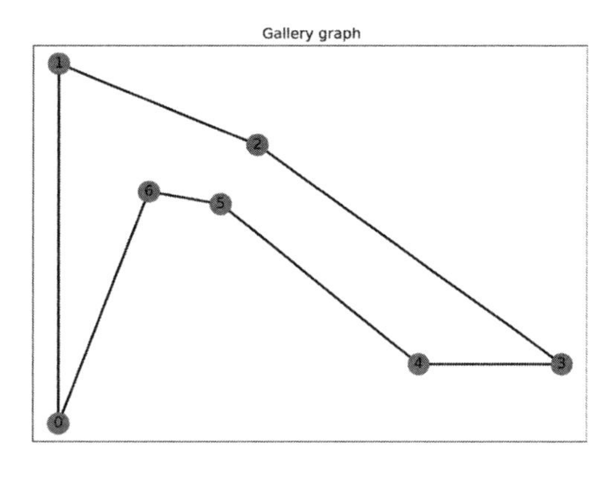

Figure 11-3: Representing the gallery with a polygon and graph

NOTE *The code to generate Figure 11-3 is in the 7th cell of the* AGP_basics.ipynb *notebook.*

On the left you see the polygon; the gray area represents the interior of the gallery. On the right you see the translation of the shape into a simple NetworkX graph. There are a number of ways you could convert the information in a `Polygon` into a graph, but often the simplest option is to iterate over the vertices of the `Polygon`'s exterior, as in Listing 11-1.

```
from shapely.geometry import Polygon, Point
import networkx as nx
import triangle as tr
❶ gallery_poly = Polygon([
    (0, 0), (0, 2), (0.55, 1.55), (1.4, 0.33),
    (1, 0.33), (0.45, 1.22), (0.25, 1.29)
    ])
gallery_coords = gallery_poly.exterior.coords[:-1]
G = nx.Graph()
❷ G.add_node( 0, coords=gallery_coords[0])
❸ pos = [list(gallery_coords[0])]
❹ for i in range(1, len(gallery_coords)):
    p = gallery_coords[i]
    pos.append(p)
  ❺ G.add_edge(i-1, i)
  ❻ G.nodes[i]["coords"] = p
    if i == len(gallery_coords)-1:
      ❼ G.add_edge(i, 0)
```

Listing 11-1: Creating the gallery representations

In Listing 11-1, we first define the gallery `Polygon` ❶ by passing in the list of vertices as before. Then we add the first node, which represents the vertices at index 0 in the coordinates list ❷. Each node in the graph is keyed by its index in the vertex list to help keep the two representations logically tied together. The first point serves as an anchor for the shape and makes it easier to create the edges—we'll do this momentarily. Before we do so, however, we have to add the coordinates to a position list called pos ❸, which will make it easier to display the graph so that it matches the shape of the polygon. Next, we loop over the remaining vertices ❹ to finish filling in the graph. For each remaining vertex, we add an edge between the previously defined node at i-1 and the current node at i ❺. Adding the edge creates the node $V_{(i)}$ and adds the edge $E_{(i-1,i)}$ in one line. Storing the coordinate information for the vertex as node metadata can serve as a more flexible alternative to the pos list. We'll do so by passing the coordinates in the coords parameter ❻. Finally, if i is equal to the last index, it's time to create the final closing edge between the last node defined and the anchor node at index 0 ❼.

Now that we have a way to generate the two representations of the gallery, let's go over the process of solving the art gallery problem in detail.

Securing the Gallery

We've already covered the start of the art gallery problem (AGP) algorithm, which creates the two base data structures we'll be working with. The next step in the process is to tessellate the geometric shape into triangles and add the resulting edges to the graph representations. Remember, any changes that you make to one representation need to be done to the other as well, to maintain their logical equivalence. Rather than using the Shapely triangulate function as we did in Chapter 7, we're going to use a purpose-built wrapper library called Triangle. The underlying application is a C-based program, also called Triangle, which was created by Professor Jonathan Shewchuk.[3] There are several reasons for choosing Triangle over Shapely. First, the Shapely version performs what's known as a Delaunay triangulation, which in its pure form doesn't respect edges; on the other hand, the Triangle library performs a *constrained* Delaunay triangulation, which can respect boundaries. Listing 11-2 shows the code to tessellate the gallery polygon.

```
tri_dict = {
    "vertices": gallery_poly.exterior.coords[:-1],
    "segments": list(G.edges())
}
triangulated = triangulate(tri_dict, "pe")
```

Listing 11-2: Performing the triangular tessellation with the Triangle library

The triangulate function expects a dictionary with two required keys. The vertices key holds the coordinates for the exterior of the shape. The segments key holds the edges that should be enforced while performing the tessellation. We pass the dictionary as the first argument to the triangulate function. The second optional argument is a string containing the settings to pass to the underlying application. There are a large number of configuration flags, and you can pass multiple arguments in the string. The p flag tells the library to treat the shape as a *planar straight-line graph*. The e flag tells the library to return the edge list as part of the result. Having the list included saves us a step when updating the graph representation, as we can simply compare the edges in the result to the edges in the graph and add any that are missing.

The `triangulate` function returns a dictionary. It includes a list called `vertices` that sets the ID for each vertex based on its position in this list. We'll need this when decoding the rest of the outputs. The other two keys we're interested in at the moment are the `triangles` key, which contains a list of triplets representing the three points making up each triangle, and the `edges` key, which contains a list of all the edges resulting from the triangulation. Both the `triangles` and `edges` use the node ID to denote vertices, so an entry in the triangle table like `[6, 1, 0]` means the sixth, first, and zeroth vertices form a triangle. If you look at the graph representation on the right side of Figure 11-3 again, you'll see that adding an edge from node 6 to node 1 ($E_{(6,1)}$) does indeed form a triangle with node 0. The `edges` list contains all of the original edges plus all the additional edges needed to form the triangles (like the edge $E_{(6,1)}$ just mentioned). You can use either the `edges` list or the `triangles` list to update the graph representation. The Jupyter notebook has an example of using the `triangles` list, but I chose the `edges` list for the code in Listing 11-3 because it's more succinct.

```
G2 = G.copy()
for e in triangulated["edges"]:
    if list(e) not in list(G2.edges()):
        G2.add_edge(e[0], e[1])
```

Listing 11-3: Updating the gallery representations

First, we create a copy of the original graph *G* to work on. We loop over the `edges` list in the triangulated result. For each edge, we check if it's already in the edge list for G_2. If it's not, we add it. The result is a triangulated representation of the gallery in both a geometric and graph representation, as in Figure 11-4.

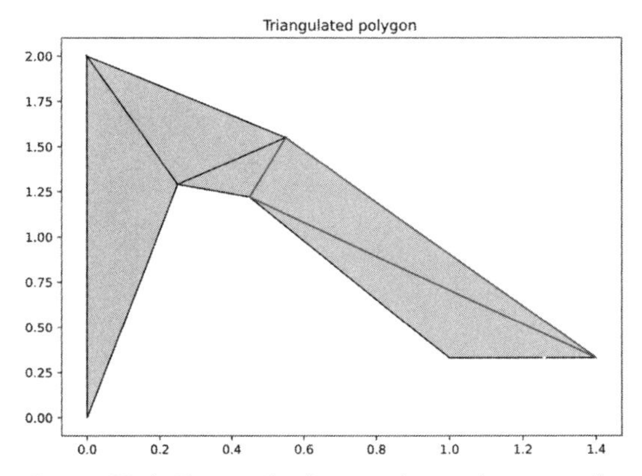

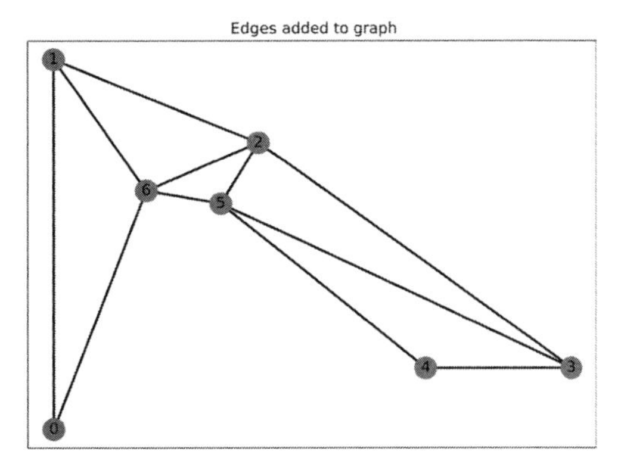

Figure 11-4: The result of triangulating the example gallery

NOTE *The code to generate Figure 11-4 is in the 9th cell of the* AGP_basics.ipynb *notebook.*

You can see that the division line segments added to the graph on the left-hand plot match the new edges added to the graph on the right-hand plot. The set of vertices and segments resulting from the tessellations is technically referred to as a *mesh*.

We're now ready to color the graph using the `greedy_color` function, which you'll need to import from the `networkx.algorithms.coloring` library like so:

```
from networkx.algorithms.coloring import greedy_color
gallery_coloring = greedy_color(G2)
```

The response from the function is a dictionary keyed off the node identifier. The value represents a numeric index for the color group the node belongs to. Figure 11-5 shows the solution for the example gallery.

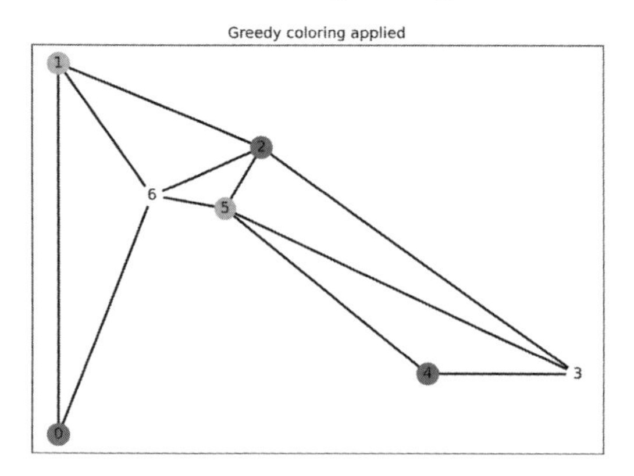

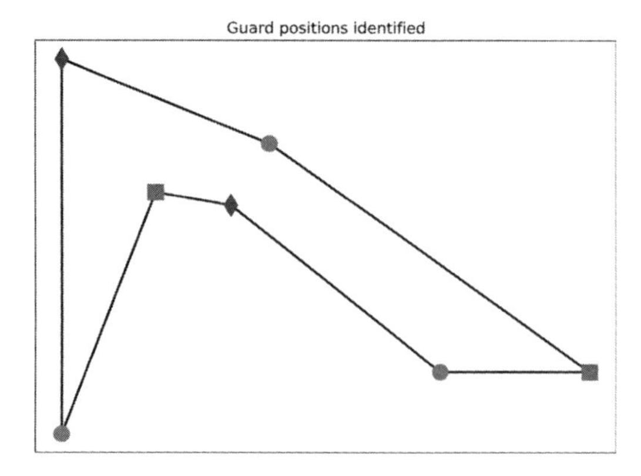

Figure 11-5: Result of greedy coloring the example gallery

NOTE *The code to generate Figure 11-5 is in the 12th cell of the* AGP_basics.ipynb *notebook.*

On the left you can see the solution found by the coloring algorithm. For the right side of the plot, I've removed the triangulation edges because we no longer need them once the coloring is done. I've also represented the color groups using different shapes—squares, diamonds, and circles—to make them easier to distinguish. You can think of each group as a potential deployment plan for guards. As you can see, not all deployments have the same number of guard positions, though. The circle group has three nodes in it while the others only have two; this means positioning guards at the locations marked by the circle nodes would require you to deploy an additional guard to cover all the walls.

You can compare the deployments that are tied for the lowest number of nodes to see what practical advantages and disadvantages each offers. For example, the square and diamond groups both require only two guards, but the diamond deployment places the guards closer to each other than the square deployment. While this may seem like small difference, in the real world having two guards in closer proximity allows them to support each other more efficiently; therefore, assuming these deployments were my only options, I would choose the diamond deployment.

Now that you've seen the basic solution concept applied to the example gallery, it's time to start refining the process and making the solution more practical for real-world use. The rest of this chapter focuses on refining the theoretical approach you've just seen to address some of the practical concerns mentioned earlier.

Mapping Guard Coverage

So far, we've ignored the scale of the gallery. The original problem assumes guards have perfect, infinite vision that isn't affected by lighting, distance, or crowds. But in a real-world scenario, the deployment needs to account for these factors with some type of maximum coverage threshold per guard location. If we say the scale for the example gallery is 1:300 m (meaning one unit on the graph is equal to 300 meters), the length of the edge $E_{(0,1)}$ is 600 m $E_{(0,1)} = 2 \times 300 \text{ m} = 600 \text{ m}$ (this is about 1,968.5 feet). The area of the gallery, then, is about 241 m^2 (2,594 square feet).

Using only two guards for a floor plan of that size is likely to leave gaps in the coverage. Add in the other environmental factors that affect visibility,[4] such as elevation (created by a slope in the floors) and lighting (often dimmed in portions of a gallery for dramatic effect), and it's clear two guards would be woefully inadequate to secure the gallery. Therefore, we need to improve on the simplifications made in the theoretical model. We'll do so by acknowledging that guards are really only able to guard so much area at a time. This is my second major reason for choosing the Triangle library: it supports the idea of a *maximum area threshold*, whereas the Shapely version of the function doesn't. The maximum area threshold sets the largest area of any triangle created during the tessellation.

If you consider each triangle as a zone that needs to be guarded, you can assign them to guard positions to create an area of responsibility (AOR) map that shows which positions are responsible for which zones as well as the overall coverage distribution (we saw an AOR map in Chapter 9 with the fire station example). The smaller the area for each zone triangle, the closer together the triangles will be clustered, and therefore more triangles will be needed to tessellate the whole gallery. More triangles mean more points in the tessellation, which translates to more guard positions. It also means more color (or shape) groups are needed to represent them.

To simplify the math, let's say each zone triangle should be a maximum area of 30 m^2 (close to 323 square feet), which scales down to an area of 0.1 unit per triangle. We can tell Triangle to tessellate the gallery into triangles with a maximum area of 30 m^2 by adding the `a` flag to the arguments string followed by the scaled maximum area 0.1:

```
triangulated = tr.triangulate(tri_dict, "pea0.1")
```

To do the tessellation, the `triangulate` function will add the points necessary (often called *Steiner points*) for the geometry to split triangles larger than the maximum area until they're all below the threshold. Figure 11-6 shows the result of the tessellation.

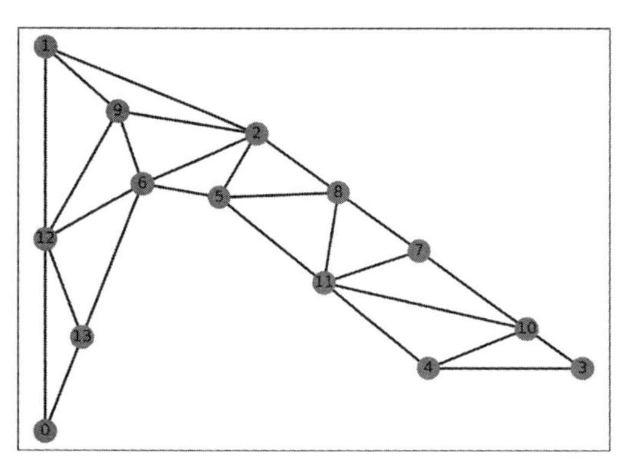

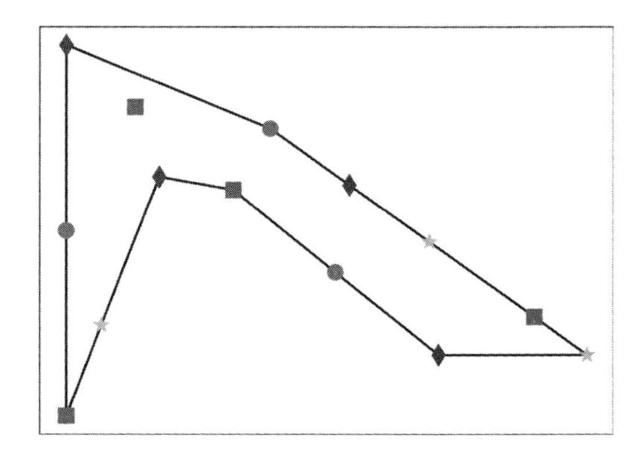

Figure 11-6: Maximum zone area 30 m^2 solutions

NOTE *The code to generate Figure 11-6 is in the 16th cell of the* AGP_basics.ipynb *notebook.*

On the left you can see the graph representation of the mesh. On the right you can see the different potential guard deployments. Since each node is a member of multiple triangles, the guard at that position will be responsible for all the triangular slices the node is a member of. Additionally, if there are any triangles that aren't directly connected to one of the guard positions (as is the case with triangle `[3, 4, 10]` in the lower right of the gallery), we assign that triangle to the closest guard position to generate the AOR map. Solving the graph coloring required the algorithm to add a fourth group, represented by the star nodes. Remember, you can think of the additional groups as more potential deployment options and the additional points within a deployment as more guards being added to that deployment. In this example, the circle deployment and the star deployments both have three positions to cover, while the square and diamond deployments both require four. By the same logic as before, I'd pick the circle deployment because it places the guard positions closer together, whereas the star deployment leaves one position relatively isolated. Figure 11-7 shows the AOR map generated from the circle deployment.

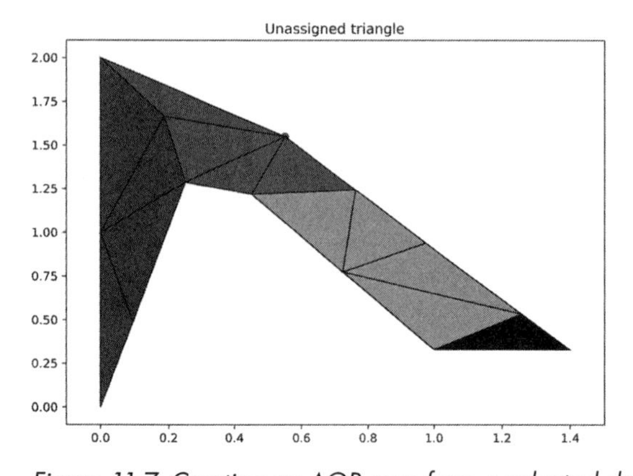

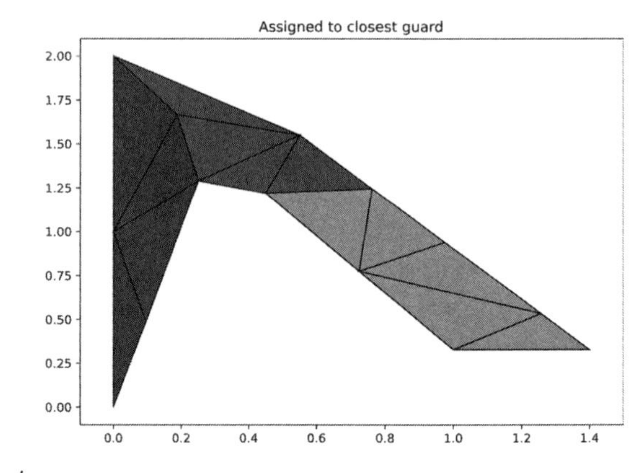

Figure 11-7: Creating an AOR map from a selected deployment

NOTE *The code to generate Figure 11-7 is in the 19th cell of the* AGP_basics.ipynb *notebook.*

The left side of the figure shows the default assignment of triangles to guards. The black triangle is the uncovered triangle mentioned previously. On the right side of the figure, you can see the result of assigning the triangle to be covered by the closest guard position. This does create a slight imbalance in the deployment. The light gray AOR contains five triangles (and therefore more square footage), while the other two AORs have only four.

Listing 11-4 shows the function for assigning triangles to guards based on the deployment group:

```
def assign_triangles(g, triangulated, group_id):
  ❶ guard_nodes = [n for n in g.nodes() if g.nodes[n]["group"] == group_id]
  ❷ triangles = {k:[] for k in guard_nodes}
    for i in range(len(triangulated["triangles"])):
      ❸ t = triangulated["triangles"][i]
      ❹ t_poly = Polygon([g.nodes[p]["coords"] for p in t])

        # If triangle touches a guard directly at any point:
      ❺ if t[0] in guard_nodes:
            triangles[t[0]].append(t_poly)
        elif t[1] in guard_nodes:
            triangles[t[1]].append(t_poly)
        elif t[2] in guard_nodes:
            triangles[t[2]].append(t_poly)
        else:
          ❻ dists = {
                k: t_poly.distance(
                     Point(g.nodes[k]["coords"])
                   ) for k in guard_nodes}
          ❼ close = min(dists, key=dists.get)
          ❽ triangles[close].append(t_poly)
  ❾ return triangles
```

Listing 11-4: Assigning triangles to guard nodes

The `assign_triangles` function takes in a copy of the triangulated graph, `g`; the result of the `triangulate` function, `triangulated`; and finally, the ID for the deployment of interest, `group_id`. We begin by collecting the nodes that represent the guard positions into a list called `guard_nodes` based on the provided group ID ❶. We then create a dictionary to hold the output of the function until the `return` statement. The keys will be the node identifiers and the values will be an empty list that will eventually contain all triangles (as `Polygon` objects) that are assigned to that node ❷.

To begin filling the `triangles` lists, we loop over all the triangles in `triangulated`. Each of these triangles represents the collection of graph nodes that form the triangle ❸. We turn the nodes into a `Polygon` by looking up the coordinates for the triangle using the provided graph `g` ❹. The next step is to check if the triangle nodes already contain one of the guard positions. If they do, then one of the three points will be in the `guard_nodes` list ❺. If the triangle isn't directly connected to a guard, we determine which guard it's closest to.

Next, we create a dictionary called `dists` that is keyed off of the guard node's ID. The value for each key will be the result of the Shapely function `t_poly.distance`, which measures the minimum distance between the triangle polygon and a `Point` object representing the guard station ❻. We then find the minimum entry in the `dists` dictionary using the `min` function. Passing in the dictionary key list in the `key` parameter tells the function to return the key that has the lowest value instead of the value itself ❼. We use this key to assign the polygon to the proper list in `triangles` ❽. Finally, we return the `triangles` dictionary to conclude the function ❾.

We can determine the exact area of each AOR by summing the area of all the triangles associated with it, as in Listing 11-5.

```
for k in triangles.keys():
    area = sum([(t.area * 300) for t in triangles[k]])
    print("Position %d covers %.2fm2" % (k, area))
```

Listing 11-5: Summing AOR areas

For each key in the resulting `triangles` dictionary, we do a `sum` operation on the list of triangle areas associated with it. Notice we adjust the scale by multiplying each area by the scaling factor (in this case `300`). The result of the code for the example should be:

```
Position 2 covers 60.41m2
Position 11 covers 98.68m2
Position 12 covers 82.34m2
```

One important note about the triangle assignment function: it's really only a heuristic. Depending on the shape of the gallery and the location of the different edges and vertices, it's possible to assign a triangle to a guard that is "closer" (as measured by minimum straight-line distance, or as the crow flies) to one guard position but logically should be assigned to another guard. You can improve the function by taking into account whether the straight-line distance measured by Shapely intersects the body of the gallery; if it does, this indicates that the measurement traveled through walls and is therefore shorter than it would be in reality (unless your guards can phase through solid objects like concrete walls). You could then change the function so that it assigns the triangles to the guard position with the shortest distance that doesn't intersect with any walls.

Another option is to change the deployment to add another guard position that covers the uncovered triangle. Looking closely at the graph representation on the right side of Figure 11-6, you can see the star at node position 3 also could have been a circle. Sometimes there will be multiple possible solutions, and the coloring algorithm has to settle on the one it feels is optimal (which means it attempts to evenly distribute color groups as much as possible). Rather than just assigning the uncovered triangle to a guard position, you might opt to change the star node to another circle node; this maintains the validity of the solution but leaves every triangle covered by the deployment. On the other hand, it also means adding another guard to the deployment, making it less optimal in that sense. A future improvement for your application may involve letting users select the assignment strategy during the AOR map creation based on their specific use.

Defining Obstructed Areas

Now that we've covered the algorithm, how it deals with scaling, and basic AOR coverage maps, it's time to deal with complex polygons. As you might recall from Chapter 7, these are polygons that have holes representing areas of the floor plan that can't be accessed or that obstruct visibility (a giant column of granite, for example).

Luckily, dealing with holes inside the Triangle library is fairly easy. For each hole we want to define, we pass the `triangulate` function a point inside the hole (any point will do). The algorithm then removes triangles until it hits a vertex defined in the `segments` portion of the dictionary. Be aware that it's possible to accidentally remove all the triangles if you improperly enclose the hole in segments. If this happens, you won't receive anything back in the `triangles` key, so you might want to update your implementation to handle this condition with some form of validity check.

To get the point for a hole, you can use the Shapely `representative_point` function, which returns a `Point` object guaranteed to be inside the boundaries of the shape the function was called on. The great thing about using the representative point is that, because Shapely doesn't care whether the point is centered in the shape, it can compute it quickly. As long as the point falls within the given shape it is considered representative, which works great for Triangle since it needs to know only where to begin removing triangles from.

Listing 11-6 shows the code to create the tessellation for the complex polygon.

```
ext_3 = [(0.0, 0.0), (0.0, 3.0), (3.0, 3.0), (3.0, 0.0)]
int_3 = [(1.0, 1.0), (1.0, 2.0), (2.0, 2.0), (2.0, 1.0)]
verts = ext_3 + int_3
❶ hole_p = Polygon(int_3).representative_point()
❷ hp1 = [list(v)[0] for v in list(hole_p.xy)]
segs = [(0,1), (1,2), (2,3), (3,0), (4,5), (5,6), (6,7), (7,4)]
sq_tri_dict = {
    "vertices": verts,
    "segments": segs,
  ❸ "holes":[hp1]
}
triangulated = tr.triangulate(sq_tri_dict, "pe")
```

Listing 11-6: Tessellating the complex polygon

We start by defining the exterior coordinates for the gallery polygon into a list named `ext_3`. In this case the points form a large square-shaped gallery. Next, we define the coordinates for the hole vertices in a list named `int_3`. These points form a smaller square hole placed directly in the center of the larger square exterior. We then create the list of vertices in a variable named `verts`, which is a concatenated list of all exterior and interior coordinates in the shape. We can calculate the representative point for the hole by casting the `int_3` list as its own polygon and calling the `representative_point` function mentioned previously.

The `hole_p` variable now holds a `Point` whose `x` and `y` values can be used to identify the hole region to the Triangle library ❶. Because Triangle doesn't work directly with `Point` objects, we must extract the coordinate information into a list ❷. Next, we create the `segs` list, which contains the list of edges that must be respected by the tessellation algorithm. Rather than using coordinates, the edges use the indices of the two vertices as the start and end of each segment. In more complex floor plans, it can be easier to take the segment list directly from the edges in the graph representation. Just be sure that all the edges and nodes are entered correctly, or you'll get some unexpected results.

Finally, we can construct the parameter dictionary for calling the `triangulate` function. The only change from the previous parameter dictionary is the addition of the `holes` key ❸, which holds a list of coordinates that represent a point inside a hole to be removed. When dealing with multiple holes, you need to calculate a representative point for each and add it to the holes list in the dictionary.

Figure 11-8 shows the steps performed on the example complex polygon gallery.

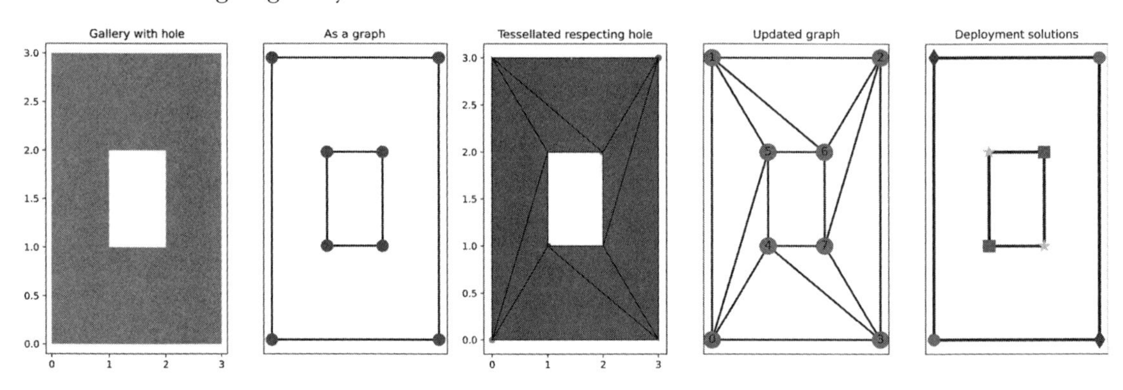

Figure 11-8: Applying the AGP algorithm to a complex polygon

NOTE *The code to generate Figure 11-8 is in the 25th cell of the* AGP_basics.ipynb *notebook.*

From left to right you can see the entire solution process applied to the gallery. The white square in the leftmost plot represents the hole in the middle of the gallery; the darker gray area represents the usable floor plan. In the second plot, you can see the result of converting the geometry into a graph. Notice that the hole section is entirely surrounded by edges. These edges are boundaries passed into the Triangle's `triangulate` function in the `segments` key (along with the exterior edges, as before). The third image shows the result of tessellating the shape with respect to the hole. The `triangulate` function begins by creating the tessellation without respecting the holes and then starts removing triangles, beginning with any triangle that contains a point in the holes list; it continues removing adjacent triangles until it reaches one of the edges in the segments list. I didn't apply an area constraint, so the result is a triangulation with the fewest number of triangles.

The fourth plot shows the updated graph, which is ready for color solving to be applied. As you can see, none of the edges in the second or third plots cross into the hole space; this indicates Triangle has respected our request to ignore that portion of the space when generating the mesh. The final plot on the right shows the result of assigning the deployment groups based on the coloring solution. There are four potential deployments, all of which consist of two guard stations. If the hole weren't present, you would theoretically need only one guard for the whole gallery.

Prioritizing Guard Coverage Areas

Next, we'll refine our model to address another implicit assumption we've made: that all areas of the gallery are equally important and therefore the guard deployments should be evenly distributed. For example, a bank may weigh the monitoring of private offices lower than that of the lobby and therefore require different resources to protect the different spaces. To address this, we'll refine our initial mesh to respect the importance (or weight) of each region of the floor plan. The underlying Triangle program supports multiple maximum areas (read in from custom data files), which allows us to get creative with how we define different AORs. To pull the example into our world of art gallery security, perhaps you believe the gallery floor needs fewer guards than the lecture hall, where people tend to gather in larger groups. You can see the result of assigning regions applied to the example square gallery in Figure 11-9.

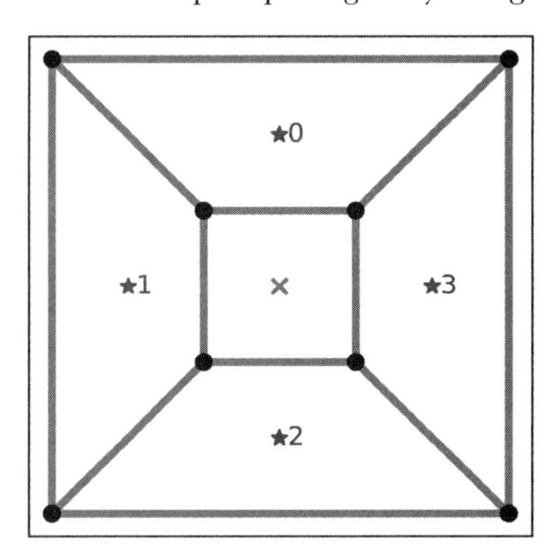

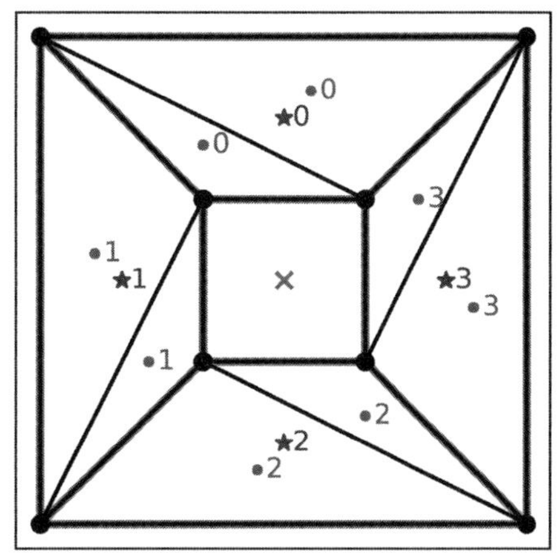

Figure 11-9: The gallery when region assignment is added

NOTE *The code to generate Figure 11-9 is in the 28th cell of the* AGP_basics.ipynb *notebook.*

On the left of the figure you can see the four numbered regions I defined for the square gallery example. The center of each region is marked by a star as well as its region number. These regions are segment-bounded portions of the overall gallery polygon, so they can be defined using any point within the segment bounds (similar to the way holes are defined). In this case, I added four theoretical segments (the diagonal separators) to the list of segment bounds already defined to enclose the regions. They're theoretical because, unlike the other segments that represent walls and other physical structures, the region segments represent logical borders between areas, not physical ones. Bear in mind that Triangle doesn't know the difference, so it's up to us to track which segments belong to which class. We can do so by taking advantage of edge attributes stored in the graph representation.

To let Triangle know that we have regions of interest, we also need to update the input dictionary we pass to the wrapper to include the `regions` key, as shown in Listing 11-7.

```
sq_tri_dict = {
    --snip--
    "regions":[
        [1.5, 2.5, 0. , 0. ],
        [0.5, 1.5, 1. , 0. ],
        [1.5, 0.5, 2. , 0. ],
        [2.5, 1.5, 3. , 0. ]]
}
```

Listing 11-7: Adding the region definitions for tessellation

The value for the `regions` key is a nested list. Each entry contains x- and y-coordinates representing a point within the associated region. During the initial tessellation, triangles within the bounding segments found around each region's representative point are assigned the value in the third position. The fourth position in the list, and all subsequent ones, can contain a numeric region attribute (such as a modified visibility value to mark regions with softer lighting). The values of any region attributes you include will be inherited by the triangles within the region.

Once the regions are defined, we can perform the initial tessellation. We'll tell Triangle to include the region information in the result by adding the `A` flag to the `triangulate` function call:

```
triangulated = tr.triangulate(sq_tri_dict, "peA")
```

The resulting mesh is shown on the right side of Figure 11-9. The dots with the numbers inside each triangle show which region identifier has been assigned to it. So far, the process is identical to the previous one, so the mesh remains unchanged.

The next step is to save the result in a special group of files; these are essentially the same data in the dictionary written out to flat text files. We need to do this so that we can have Triangle reload the files during the refinement stage. Unfortunately, the wrapper library doesn't include any functions to save, so I had to write my own based off the file specification in the underlying program's documentation. The project code has all the functions necessary to create the expected files, but they're long and fairly boring, so I'll skip detailing them here. All you need to do is import the DataSaver class I've included with the chapter code, as shown in Listing 11-8.

```
from filemanager import DataSaver
saver = DataSaver(triangulated, "square", 1, "/myproject/")
saver.set_region_areas([-1,0.1,-1,0.05])
saver.save_project()
```

Listing 11-8: Saving the tessellation result to Triangle project files

We start by defining the DataSaver instance, passing it the triangulated dictionary in the first parameter and the project name in the second. You can also optionally pass in a version number and directory. If you don't pass these in, the version number will automatically be set to 1 and the directory will be the current working directory of the script. The set_region_areas function stores the list of maximum areas, and the index in the list is the same as the region index. If you pass in a list that is shorter than the number of regions, the remaining regions are treated as though they had no area constraint.

We need to set the region areas before we save the project using save_project so that the saver knows how to mark each triangle during the file creation. Once save_file function has completed, you'll have a list of files in the directory you specified (or the current working directory, as I mentioned). Each file is named like so: *<project name>_<version number>.<part>* where *<part>* is one of the file types expected by Triangle (with a *.node*, *.ele*, *.area*, or *.poly* file extension). Therefore, the code in Listing 11-8 will create a file named */myproject/square_1.area* (among others). The *.area* file is of particular interest because it contains the maximum area for each triangle, which we just set by virtue of its region association.

Next we need to refine the mesh by reloading the saved data and performing another tessellation:

```
reload = tr.load("/myproject/", "square_1")
refined = tr.triangulate(reload, "ra")
```

We call the load function from the wrapper library and pass in the directory where the project files were stored, as well as the project name, including the version. These inputs find and load the data from the associated files created in Listing 11-5. Finally, we create the refined mesh by calling the triangulate function once again, this time on the loaded data. The option string ra tells Triangle to refine a previously generated mesh (the r flag) and to refine the mesh using constrained triangle areas (the a flag). Because Triangle is refining a mesh, and we've asked it to constrain the area, it will attempt to locate the project's *.area* file and use that information while creating the refined mesh.

During the refinement step, each triangle in the original tessellation is compared against the area defined in the *.area* file. If the area is larger than the maximum defined, the algorithm splits the triangle into smaller ones. You can see the result of the region areas being applied to the tessellation of the square gallery in Figure 11-10.

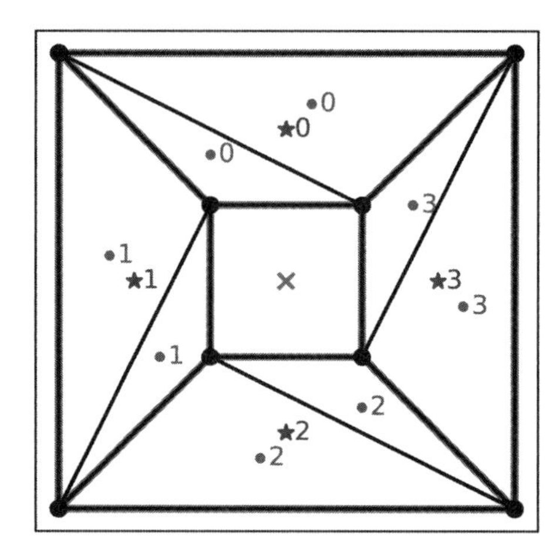

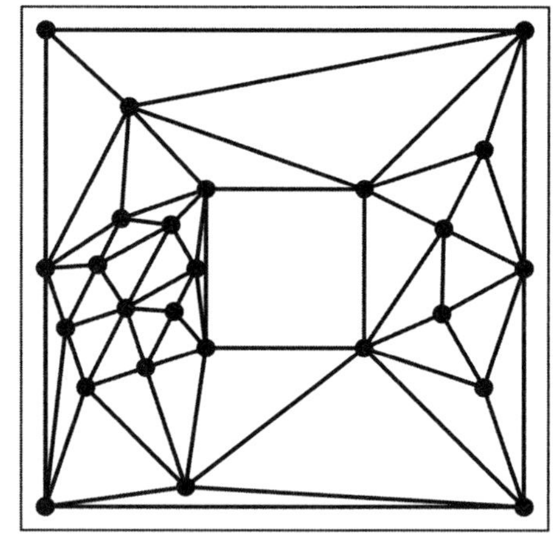

Figure 11-10: Creating a multiregion tessellation

NOTE *The code to generate Figure 11-10 is in the 31st cell of the* AGP_basics.ipynb *notebook.*

On the left, you can see the region assignment. The areas passed in to the `DataSaver` class were `[-1,0.1,-1,0.3]`, which can be interpreted as: region 0 has no maximum area constraint (marked by any negative value); region 1 has a maximum area constraint of 0.1; region 2 also has no constraint applied; and finally region 3 has a maximum area constraint of 0.3. On the right side, region 1 has been broken up into many small triangles, while region 3 has been divided up into slightly larger triangles. The result is more guard positions in region 1 than region 3, and increased coverage in both compared to the other two zones with no constraints applied.

We can save the refined mesh using another instance of the `DataSaver` class:

```
saver = DataSaver(refined, "square", 2)
wsaver.save_project()
```

We pass in the `refined` mesh result, project name, and version once again. It's standard practice to increment the version for each subsequent mesh you refine, so this refined mesh would be the second version of the square project. We don't need to define the region areas this time unless we want to perform any further refinements.

Of course, it'd be nice if you could just add as many guards as you wanted to the deployment. Unfortunately, we rarely have the budget for more than some number of guards per shift or for some fixed number of sensor devices we can deploy per floor. We can tell Triangle the maximum number of Steiner points it can add to achieve the tessellation with the `S` flag. For example, let's say we can only afford to deploy three extra guards to the square gallery. Figure 11-11 shows the result of adding this constraint during the refinement step.

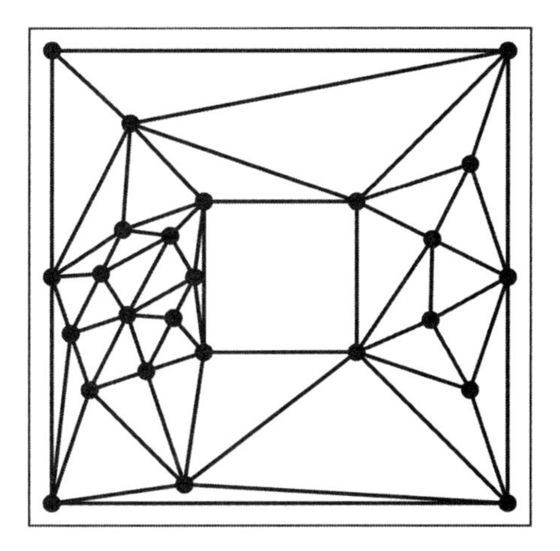
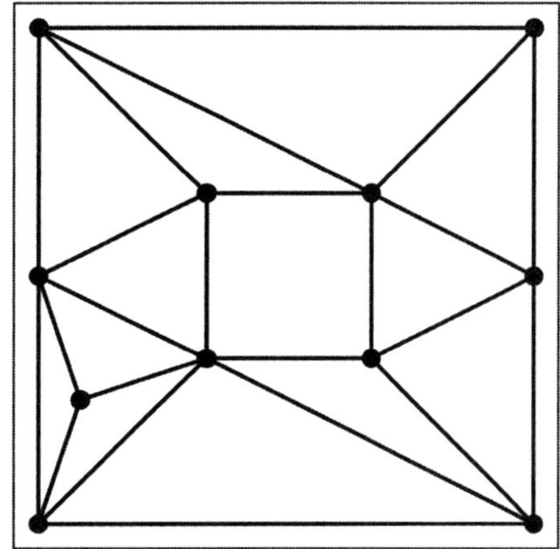

Figure 11-11: Refined mesh with three guards

NOTE *The code to generate Figure 11-11 is in the 34th cell of the* AGP_basics.ipynb *notebook.*

On the left, you can see the mesh we generated previously, with no restriction on the addition of guard points. On the right, you can see the mesh generated from the refinement where I restricted the output to three additional guard positions. Bear in mind that I didn't change the maximum area constraint I defined earlier, which clearly the mesh on the right doesn't achieve. Once Triangle runs out of Steiner points it can add, it can't divide the larger triangles anymore and so it stops. If you're asked to divide the gallery into small zones and then cover those zones with only three additional guards, you can use this result to show that it's a mathematical impossibility to accomplish both goals. Hopefully this result gets you the budget increase you need for those additional guards!

Mapping Security Camera Field of View

The next feature to address is adding the field of view and effective range parameters. When discussing human guards, the field of view and effective range are very different from person to person so assigning these parameters is a bit fuzzy. Intuition and educated estimates are your best friend in these cases. However, when you think of guard positions as electronic devices, such as cameras or motion detectors, the data is easier to locate. For example, I searched Google for "security camera data sheet" and selected the first model I saw: Ocuity model number HMNC100 from NetGear. After reviewing the technical data for the model (available from the manufacturer's website), I found that the listed field of view was 107 degrees and the effective range was 7 m in total darkness (thanks to built-in infrared lighting). The distance during normal operational lighting is often not listed because the answer depends on how much detail you need to be able to distinguish, the focal length of the camera (which may or may not be listed in the user documents), and the number of pixels used to encode the information (often given as a megapixel rating, where 1MP is equal to one million pixels). To get a good number for effective distance, the best method is often to just test the camera under the conditions of interest.

We'll assume the normal effective range *r(d)* is 20 times greater than the night range, $r(n) = 7$ m ($r(n) \times 20 = 140$ m or about 460 feet). Since we're approximating things, I'm also going to change the field of view Δ from 107 to 104 because 107 is prime and therefore can't be divided up into equal segments; this makes computing the points representing the edges of the field of view harder. On the other hand, 104 can be divided by several factors, including 2, which is also helpful for keeping the math simple. I'll define the starting angle $\Theta_{(0)}$ for each of the four internal guard nodes `[4, 5, 6, 7]` as `[180, 134, 45, 351]`, respectively. We can define the positive peripheral angle as $\theta_{(p)} = \theta_{(0)} + \frac{1}{2}\Delta$ and the negative peripheral angle as $\theta_{(n)} = \theta_{(0)} - \frac{1}{2}\Delta$.

Finally, we can create some number of intermediary angles between the two peripheral angles by dividing the field of view into eight segments of 13 degrees each:

$$\delta = \Delta / 8 = 13$$

We can then use the `cos` and `sin` functions for each of the angles:

$$A := [\theta_{(p)} - i\delta] \, \forall i \in \mathbb{Z}_{0-8}$$

Figure 11-12 shows the approximated coverage if you positioned one of these cameras at each of the four internal guard positions from the square example in Figure 11-8, using the starting angles just mentioned.

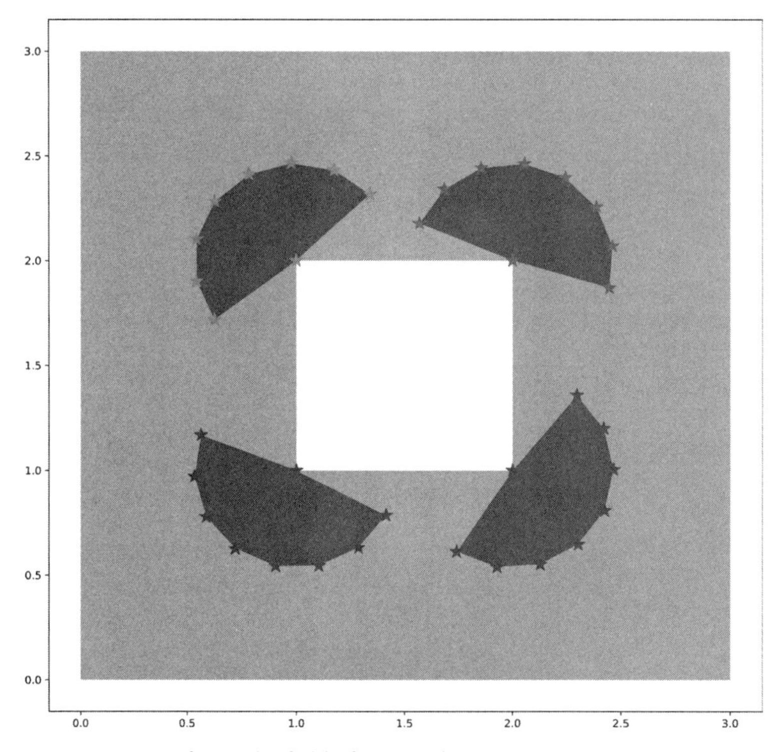

Figure 11-12: Defining the field of view polygons

NOTE *The code to generate Figure 11-12 is in the 26th cell of the* AGP_basics.ipynb *notebook.*

Figure 11-12 shows the approximate field of view coverage resulting from placing one of those cameras at each of the corners of the internal hole. As you can see, there are clearly large areas the cameras aren't able to adequately cover. Furthermore, you can see someone could theoretically approach each camera position without ever entering the field of vision of any camera. These are examples of the blind spots mentioned earlier in the chapter. As a defender, you'd want to add more coverage to prevent these gaps. As an attacker, you could consider all of the uncovered assets as potentially good targets.

One last note about field of view mapping. I've assumed that the cameras are pointed and then left stationary. Several high-end cameras include a device called a gimbal that allows a remote operator to move the camera as needed. Some mid-level cameras also include the ability to automatically sweep back and forth across an area. These types of cameras are becoming less popular these days, but they're still good to be aware of; not all field of view maps are going to be static. Be sure to look out for features like these when you're analyzing the data sheets for devices; you may need to update your implementation to produce the full range of vision for these cases.

Summary

You now have all the code necessary to solve the core of the art gallery problem, including methods for handling real-world constraints, such as limited budgets and effective range for sensors. We've discussed the power of the Triangle library and combined it with Shapely and NetworkX to model the problem with both a geometric and graph representation. Perhaps most important of all, you should now be comfortable explaining the AGP theory and discussing the practical constraints applied within the application. There's a large body of research you can tap into to continue developing your understanding of the problem. I didn't even touch on how the greedy color algorithm functions under the hood, for example.[5] Node coloring in general is a great graph theory topic with many applications outside the AGP, but for our purposes, studying node-coloring algorithms can help you understand the solutions your application outputs (and how you can potentially modify them) more completely. For a more in-depth description of the geometric implications of different gallery layouts, check out the paper "Note on an Art Gallery Problem."[6]

To continue developing the core of the system, you can identify additional use cases that may be of interest to the users you want to serve. For example, you might want a use case to cover users who want to compare before and after deployment plans. Once you collect the use cases, you can develop the additional functions needed to support them. Additionally, you can continue to refine the functions included here. The triangle assignment function is one excellent area to improve upon using one of the options we discussed previously. Remember, you don't necessarily need to develop all of these features before taking your application to the market. Developing an MVP is all about picking the key features that make your application useful to users. Oftentimes, figuring out these key features is a matter of analyzing competitive software offerings and answering the questions, "What features do all these applications offer?" and "What features does my application offer that the others do not?"

Once you can answer those two questions, you're ready to move on to the next chapters, where we'll go from the core algorithm developed in this chapter to a full-fledged Python application. We'll finish out the project by mapping out the user interaction, adding graphics, selecting a modern processing architecture, and deciding whether to add application licensing.

12

THE MINIMUM VIABLE PRODUCT APPROACH TO SECURITY SOFTWARE DEVELOPMENT

Going beyond your current proof-of-concept code means planning how other users will interact with your program. Mapping out the user's path through the application from start to finish will allow you to decide the best ways to deliver the application, which in turn will enable your users to quickly and intuitively begin using your software. The considerations in this chapter apply to most types of applications equally well, as they deal with the process of delivering and using the application, rather than with the problem directly.

In the previous chapter we identified a few use cases that led us to define several features that the application would have to support. In this chapter, we're going to implement the still-open features: developing a graphical user interface (GUI) and saving projects. We'll begin by mapping the user's interactions with the program, using them to build the GUI. Then we'll discuss state managers and how to use process parallelism to split up the workload, optimizing our solutions for complex floor plans. Finally, we'll build the GUI and implement the save feature.

Mapping the User's Interactions

Most good software projects begin the development phase with a set of user interaction plans that describe the actual steps a user will follow in the application to achieve each use case. These plans can be rough sketches or extremely detailed wireframe mock-ups of the final application, but they all need to answer the question, "How will the user use the system to achieve the goal?" My preferred method for plotting use cases is using application state machines for the structure and Unified Modeling Language (UML) for the process. I recommend using an application like Dia or LucidChart to produce a visual layout. Figure 12-1 shows the workflow diagram I developed in LucidChart for the use case where a user wants to save their project across multiple sessions.

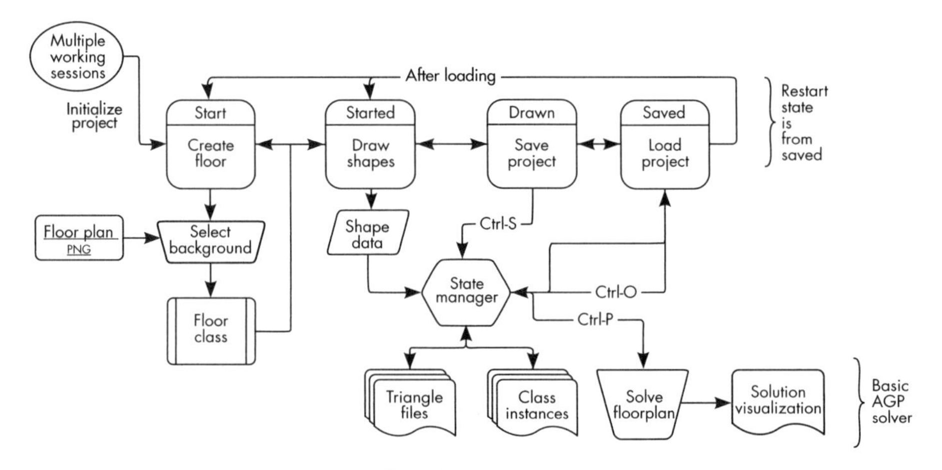

Figure 12-1: A multiple-session workflow diagram

I chose this particular use case because it encompasses the entire application flow: we create multiple floors, draw the polygon representation, save and reload project data, as well as create the final solution visualization. At the top left of Figure 12-1, you see the oval shape, which is used to represent the use case entry point. In this scenario, a user who wants to work across multiple sessions begins by initializing their project. Right now that simply means they open the program for the first time, but we'll add more to the initialization in later sections.

The rounded squares represent high-level states of the application. These states are similar to the ones we examined in Chapter 6 when we discussed finite state machines: they represent the options available at each state to transition to another state. When a user initializes their project, they're moved into the *Start* state, where they're expected to create one or more `Floor` objects that contain a background PNG file. Once the `Floor` class representing the floor plan is created, it's passed to the next state the user enters, the *Started* state.

In the Started state, the user has the option to draw the shape data for the floor plan. To do so, they click their mouse to create a series of points representing the hull of the floor plan along with points within the hull to define holes (represented by the `Obstacle` class in the application primitives). Once a user adds a point to start drawing a shape, they must either finish drawing or back out using an `undo` function. After the user draws the shapes they want, the Shapely data is sent to the state manager, the heart of the whole application. We'll cover the state manager in more detail momentarily, but for now just understand that the state manager's job is to track what the user's doing in the application and expose the proper actions and impose the proper constraints (such as finishing a shape before saving).

Once the user has drawn as much as they want for the session, they move into the *Drawn* state, where they'll save the information for the next session. To tell the state manager to record the various files and objects, the user enters the key combination CTRL-S. From this state, the user can either return to drawing—allowing incremental progress saves within a single session—or end the program and return later. When the user does return for subsequent sessions, they enter the application at the *Saved* state, which allows them to ask the session manager to load a previously saved project using the key combination CTRL-O. After loading completes, the user can add new floors (the Start state) or continue to draw on previously created floors (the Started state). Once all the necessary shapes are recorded, the user can solve the floor with the key combination CTRL-P. The result will be an image file that includes the lowest-count deployment overlaid on the background image for the floor plan.

Overall, this implementation is still somewhat basic in its feature set; this is intentional to allow you to expand the program to meet your vision for the project. Part of the art in application development lies in how you choose to organize the feature work, so I won't go into any more detail about which pieces I think should go where. As long as you lay out the features to match the use cases in a way that makes sense to you, they'll serve as a map to guide the rest of your project development.

Planning Application States

Once you've created the rest of your application state flowcharts, you can begin to pick apart the code needed to support each application state. This process is about identifying the pieces of the application that impact the most features and developing the code necessary to support those functions that cover it. Following this application-mapping process helps you reduce unnecessary code by identifying the shared pieces of an application so you can develop them with reusable classes or functions. Because it allows you to track which interactions touch which sections of the code, this process also helps you gauge the relative importance of each piece to the application's performance. For example, by examining the diagram in Figure 12-1, we can tell that we'll need a state manager class to track what's happening in the application, a floor class to represent the floor plan, and some way to support keyboard input (like the hotkey commands CTRL-S and CTRL-O). We know the state manager will need to be able to take in shape data and output files that store the important details. It will also need to be able to read in those same files and rebuild the last saved state from them.

Creating an application map also helps you organize the libraries and modules you'll need to support these requirements. We already know we'll need the Triangle and NetworkX libraries from the previous chapter. Because we'll need graphic displays, keyboard shortcuts, and mouse interactions, the PyGame library is a good choice for developing the user interface—it supports all three needs at once. Since we're using a background PNG file, we'll need something that knows how to work with this file type. We could use a specific PNG library (creatively named png) that works well with PyGame's functions, or we could use imutils or another similar library. Allow your application diagram to guide your selection of libraries and read the documentation for different options. Look for libraries like PyGame that can solve multiple requirements. Where two options appear equally suited to a task, I often choose the one that allows me to reduce the overall requirement size. That means choosing the one that's already included in the requirements library or selecting the smaller of the two packages in terms of file size.

In practical development, the number of use cases you include and the order in which you approach them will largely be dictated by external forces (such as business needs and budgets), so it's impossible for anyone to give you a generic development process to follow. Still, at a high level, developing use case diagrams and application flow diagrams will almost always help you stay on target.

Next, let's discuss an often-overlooked topic when it comes to proof-of-concept development: documenting our projects for the good of humanity.

Documenting the Application

An absolute must for any important project is good documentation. We've already got a start on our documentation with the application state diagram. When it's time to deliver this software, we can include this diagram along with other project documentation to show users our application's basic functionality. In addition to the project artifacts, we should also be documenting our source code. The code for this project uses a method known as *docstrings* to document the code both for ourselves and for future developers. Docstrings are comments added directly to your script files in a human-readable syntax. This topic is covered in great detail online (*https://peps.python.org/pep-0257*), so I won't go into it much here, but I do think it is helpful to show an example:

```
def concat_str(a, b):
    '''
    Returns the concatenation of string 'b' to string 'a'.
            Parameters:
                    a (str): A string literal
                    b (str): Another string literal
            Returns:
                    concat_str (str): String after adding 'b' to the end of 'a'
    '''
    concat_str = a + b
    return concat_str
print(concat_str.__doc__)
```

The docstring for the function is the string literal enclosed in triple quotes (`'''`) directly after the function definition. The comments cover the input and output of functions (in the example under the `Parameters` and `Returns` headers, respectively), document the intended use of class objects, and record any important notes or facts for anyone who might want to make future changes—for example, including a link to the source material for a particular algorithm the function uses. We can access the documentation for any function with a docstring using the built-in `.__doc__` attribute. The docstring syntax also allows automated programs to detect the comments and format them into prettier API documentation for general distribution.

Having good documentation allows you to quickly bring in new developers. If you were developing this application for a business, you could easily train the other developers on your team and enable them to develop extensions and improvements. Likewise, if you're building for the open source community, good documentation encourages good contributions. Even if you're building only for yourself, strong documentation practices can help refresh your memory when you pick up a project after a long time away from it.

Now that we've discussed all the planning essentials to keep our code efficient and promote adoption by other developers, we can begin the fun part of the project: developing the core of the application, the state manager.

Developing the State Manager

Extremely common in modern software, *state managers* manage all the different possible interactions in a program. For example, your web browser has to keep track of where you click and what you type, as well as everything else happening in the browser. The state of the browser will determine what those clicks and keypresses do: pressing CTRL-S on the Google home page attempts to save the page as an HTML file, while the same shortcut on the Google Docs web page saves the project to your cloud storage. This can happen only because there's a class responsible for orchestrating all the pieces: the state manager. State managers are *event driven*: when certain actions—like a pressed key or right-click—happen, they get notified and decide how to respond.

Let's take a look at some events. Listing 12-1 shows an example of how PyGame uses the `pygame.event` class to send events to your program.

```
import pygame, sys
import state_manager as state
❶ for event in pygame.event.get():
  ❷ if event.type == pygame.QUIT:
        --snip--
        sys.exit()
    elif event.type == pygame.MOUSEBUTTONDOWN:
      ❸ state.handle_click(event)
    elif event.type == pygame.KEYDOWN:
```

```
        pygame.event.clear(None)
        state.handle_keydown(event)
    elif event.type == pygame.KEYUP:
        state.handle_keyup(event)
 ❹ elif event.type != pygame.MOUSEMOTION:
        print(event.type)
```

Listing 12-1: Handling PyGame events

The pygame.event class contains a queue of all the events that happen between successive get requests. The result of the get call is a list of event objects to process that we'll loop through ❶. The order of the events will (usually) be the order in which the user performed them. For example, a user entering the CTRL-S keyboard shortcut to save a project would set off a series of events in the queue representing the user pressing CTRL, pressing the S key, then releasing the S key, and finally releasing the CTRL key.

Each event object has a type field to help you understand what occurred. There are five major event types we're concerned with. pygame.QUIT ❷ is a special event that should trigger your application's shutdown code and finally conclude the program by exiting the application cleanly (without leaving unused files on disk or resources opened). The other types also have fairly intuitive names: the pygame.KEYDOWN and pygame.KEYUP events occur when the user presses or releases a keyboard key, respectively. Similarly, the pygame .MOUSEBUTTONDOWN event indicates a user clicked one of the mouse buttons ❸. The state manager uses these events to decide when and how to transition between application states.

There are a lot of events in the PyGame event queue. The pygame .MOUSEMOTION event, for example, is triggered multiple times as the user moves the mouse pointer around the screen. You can filter undesirable events by checking that the event type is not equal (!=) to an undesired event's type ❹. During development, printing events out to the console can help you identify events that you might want to develop code for, such as additional keyboard shortcuts. This is an excellent time to use type filtering to reduce the number of output messages, as shown in Listing 12-1.

Modern applications like web browsers and operating systems can have hundreds of states or more. Furthermore, each of these states can have *substates*, different options that exist within a single state, such as a red or black background screen in the drawing state. The more complex the application becomes, the more important it is to refer back to your workflow diagrams from time to time. Development is an iterative process, and it's important to make sure you're capturing all the major states as the project progresses. Rather than detailing all the code in the state manager, I'll cover examples of the key elements that drive the interactions. For example, rather than showing you the logic for handling every keystroke, I'll show you how to handle the two types of keyboard input (KEYUP and KEYDOWN) generically. You can then dive into the specifics for each function using the documentation provided in the file *AGP_solver_API.pdf* in the project's resources folder.

The code in Listing 12-2 shows the framework for the handle_keydown function.

```
def handle_keydown(event):
    global shifted
    global controlled
  ❶ if event.key in [303, 304]:
        # Shift key depressed.
      ❷ shifted = True
        return
  ❸ if event.key == 306:
        controlled = True
        return
  ❹ if event.unicode == "z":
      ❺ undo()
```

Listing 12-2: Handling `KEYDOWN` *events in the state manager*

The function takes in the `pygame.event` object as its only parameter. It needs the object so it can determine what key was pressed and respond accordingly. There are a few ways to check the value of an event. First, by checking to see if the `event.key` attribute is in a list of values ❶, you can apply the same code block to multiple input values. These values are numeric identifiers for every key on the keyboard, also called the key's *scan code*. In this case, the two values `303` and `304` correspond to the left and right SHIFT keys, respectively. If either is pressed, it will trigger the code that sets the `shifted` variable to `True` ❷. If you care about only one key from a pair, you can compare the `event.key` parameter to a single scan code ❸. In this case, `306` corresponds to the left CTRL key. Pressing the right CTRL key won't trigger the code block that sets the `controlled` variable to `True`. Note that the `shifted` and `controlled` variables are both global variables, which means their value will persist in the application even after the function returns. This is what enables us to know if the user has entered a two-key combination like CTRL-S, which takes two calls to `handle_keydown` to achieve. PyGame also has another built-in function called `pygame.key.get_mods` for determining whether modifier keys like CTRL or SHIFT were pressed in a key combination; you should explore it to improve the code from Listing 12-2.

Determining if a specific key was pressed is sometimes easier to read in the code if you compare the `event.unicode` attribute against the string literal of the key instead of using the key's scan code. In the example, we compare the attribute to the string `"z"` ❹; if the value matches, we call the `undo` function ❺.

NOTE *The code for the* `undo` *function is in lines 418–503 of the* state_manager.py *file. It essentially tells the state manager to go back to the previous state.*

The `handle_keyup` function in Listing 12-3 is shorter because there are usually fewer keys whose release timing we care about.

```
def handle_keyup(event):
    global shifted
    global controlled
```

```
    if event.key in [303, 304]:
        shifted = False
        return
    if event.key == 306:
        controlled = False
        return
```

Listing 12-3: Handling KEYUP *events in the state manager*

The code for this function is the inverse of that in Listing 12-2. We check the `event.key` parameter to see if either SHIFT key was released by the user and, if so, set the `shifted` key to `False`. Otherwise, we check if the left CTRL key was released, in which case we set the controlled variable to `False`. The Z key fires off a one-time event, which means you don't need to worry about whether the user released it or not.

The state manager has a similar function named `handle_click` that handles mouse-click events, as shown in Listing 12-4.

```
def handle_click(event):
 ❶ clicked = check_clicked_existing_vertex(event.pos)
 ❷ in_room = check_clicked_within_room(event.pos)
 ❸ if event.button == 1:
        --snip--
      ❹ left_click(event, clicked, in_room)
    elif event.button == 3:
        --snip--
      ❺ right_click(event, clicked, in_room)
```

Listing 12-4: Handling MOUSEBUTTONDOWN *events in the state manager*

The concept here is the same as in the previous listings, but the logic has to be longer to handle determining what button was clicked, where the pointer was when the click happened, and what other objects were in the same area as the pointer when the button was clicked. The `check_clicked_existing_vertex` function ❶ compares the position of the mouse (stored in the `event.pos` parameter of mouse events) against a list of all vertices in the project. It's hard for a user to click exactly where a vertex is, so we give them a bit of room for error, called ε (epsilon). The current value for epsilon is 3 pixels. If the pointer was within epsilon (3 pixels) of a vertex, the vertex's data is returned. The `EPSILON` constant is defined on line 35 of the *state_manager.py* file in the chapter's supplemental materials, and the code for the `check_clicked_existing_vertex` is in the same file starting at line 634.

Similarly, the `check_clicked_within_room` function ❷ checks to see if the pointer was inside any of the polygon shapes. The event and the information regarding the clicked objects (if any are present) are then passed to the appropriate function based on what button was clicked. Some mice have different button configurations, and the scan codes will depend on the manufacturer and driver your computer uses to some degree. You can use `print(event)` while clicking different buttons to have PyGame identify their scan codes for you. For production applications, you should use PyGame's built-in key literals like `pygame.key.K_a` instead of scan codes for portability.

In Listing 12-4, if `event.button` is `1` ❸, it corresponds to the user clicking the left mouse button, so we call the function `left_click` ❹. If `event.button` is `3`, it corresponds to the user clicking the right mouse button, in which case we call the `right_click` function ❺. Both of these functions take in the `event` object, the `clicked` vertex, and `room` (if any exist). Each function uses these details to determine how to update the internal state using a lot of specific logical checks. For example, implementing a special delete response for a SHIFT-left-click on a vertex requires the `left_click` function to first check if the `shifted` global variable is set to `True`. If so, the state manager removes the vertex that was passed to it for the list of vertices, thus updating the internal state. If the `shifted` variable is `False`, the state manager will follow a different branch of logic depending on whether a room, vertex, or neither was clicked. As you can imagine, the logic for these functions can quickly become quite long and complex.

Handling events and managing state as we've done in the last three listings is the central concept behind the state manager code. As you expand your implementation, you'll continue to add logic to the `handle_keydown`, `handle_keyup`, and `handle_click` functions to implement all the different possible user interactions, such as drawing the rooms and adding obstructions to them (covered in the "Adding a Graphical User Interface" section shortly).

Accelerating Security with Parallel Processing

Process parallelism is a large topic with many nuances, but simply put, it means spreading the work that needs to be accomplished across multiple workers. How you accomplish this is called the *division of labor* and is a matter of some contention. For example, suppose you're a teacher and you have to grade 100 student papers, each with 20 questions. Luckily, you have four teaching assistants to help you, for a total of five workers grading papers, so you could each grade 20 papers. The benefit would be that you have five papers being graded at the same time instead of just one.

Another option would be for each of you to select four questions to grade. You take the first paper, grade your four questions, then pass it along to the next person for them to grade their four questions, and so on. In this scenario, every worker touches every paper at least once, but for a reduced amount of time. The benefit of a division of labor like this is it allows each worker to focus on the work they're best suited for. Imagine a scenario where one of the assistants is a specialist in mechanical engineering and another specializes in chemistry. By allowing each specialist to work on their area of expertise, you maximize the speed of the process by leveraging their individual capabilities.

Threading Parallelism

More formally speaking, Python has two main approaches for parallelism: threading parallelism and processor parallelism. *Threading parallelism* occurs when a main application opens up a child application that shares its resources, like memory space (*https://docs.python.org/3/library/threading.html*).

This would be similar to having a single answer key for the test that all the people grading papers can look at (imagine it's taped to a wall in the room). Each grader represents a thread and the answer key is a shared resource they all have access to. With most flavors of Python, threading isn't technically parallelism because only one thread can be executing a command at a time (controlled by the Python interpreter). This is equivalent to letting only one person grade an answer at any given time. For practical purposes, though, switching threads happens so fast that it's nearly simultaneous, so most developers (myself included) still count it.

Threads can be created in a few different ways, but one of the most popular is to create multiple threads of a single class, as shown in Listing 12-5.

```
import threading, os, Image
❶ class DisplayAGP(threading.Thread):
    open_image = None
  ❷ def set_file(self, bgd_file):
        self.open_image = bgd_file

  ❸ def run(self):
        file, ext = os.path.splitext(self.open_image)
        im = Image.open(self.open_image)
      ❹ im.show()
        return
```

Listing 12-5: Threaded parallelism for displaying concurrent images

The code defines a class named `DisplayAGP` that allows the application to display multiple images at the same time, while still allowing the main application to do other work in the background. All classes that are intended to be used in a threaded fashion need to extend the `threading.Thread` class ❶ and contain a `run` method ❸, which contains the logic to be executed within the context of the thread—in this case, opening a particular image file and displaying it with `im.show` ❹. You can add further class methods, like the `set_file` method ❷, as a means of configuring each thread prior to running it.

To display all the image files in a list creatively named `image_files` concurrently, you can use the code in Listing 12-6.

```
for fp in image_files:
    t = DisplayAGP()
    t.set_file(fp)
    t.daemon = True
    t.start()
```

Listing 12-6: Displaying images concurrently with the `DisplayAGP` class

We start by looping over the list of file locations. For each one, we create a new instance of the `DisplayAGP` class. Then we call the `set_file` method with the location of the image so each thread knows what it should display. Setting the `t.daemon` property to `True` tells the program not to wait for the result of the thread once it is started with the `start` method. Calling `t.start` actually triggers the code in the `DisplayAGP.run` method. Once all the threads have been started, your main thread is free to move on to handle other work.

NOTE *You can see how I've implemented this code in lines 602–605 in the* state_manager.py *file. There I've hardcoded an output filename. You should fix this to be flexible in your own deployment. See the Python documentation for more details on how to implement threading for different scenarios.*

Processor Parallelism

One of the major drawbacks to threading parallelism is that all the threads live in one application, which in turn lives in one section of your processor. Chances are your computer, phone, tablet, and probably even your toaster have multiple cores in their central processors. Threads load all the workers onto one core while the other cores sit idle (at least from Python's perspective), losing all the advantages of modern computer architecture. *Processor parallelism* (also called *multiprocessing*) aims to address this limitation by enabling multiple copies of an application to communicate with one another (*https://docs.python.org/3/library/multiprocessing.html*). By splitting the work across separate instances of an application, you allow your processor to divide the work among all the cores. These cores are each independent little processors that can execute instructions on the same clock cycle as the other cores, making it the truest form of parallelism available in Python. Going back to our test grading analogy, this would be like giving each grader their own copy of the answer sheet to take home and work from. The graders would each get a copy of the resources they need to complete their portion of the tests (such as the answer key and a stack of questions), but they could also use whatever additional resources they have access to at home (such as a faster computer). Each grader operates in their own environment, completely independent of the other people grading papers. Once someone finishes grading their portion of the tests, they bring their graded answers back to you (the teacher, or main process), who then compiles the individual responses into final scores.

Listing 12-7 shows how we can call a separate process to solve a multi-floor project.

```
def mp_solve_floors(floors):
 ❶ ctx = mp.get_context('fork')
 ❷ q = ctx.Queue()
   p = ctx.Queue()
   procs = []
   for i in range(0,len(floors)):
     ❸ proc = ctx.Process(target=mp_agp_solver, args=(p,q))
     ❹ proc.start()
       procs.append(proc)
   results = []
   for f in floors:
       --snip--
     ❺ p.put(work_item)
```

```
    while True:
      ❻ results.append(q.get())
    for proc in procs:
      ❼ proc.join()
    return results
```

Listing 12-7: Using multiprocessing to solve floors concurrently

NOTE *The full code for this function is in the* graph_shapes.py *file in lines 79–107.*

The `mp_` preface in the function name `mp_solve_floors` is a standard way to denote functions that are designed to be part of a multiprocessing architecture. The function expects a list of `Polygon` objects representing the floors to be solved using the AGP solver we developed in the previous chapter.

We first define the context that we'll use to create the other processes. There are a few options, and the best one depends on your use case and the underlying system. The `fork` ❶ context will create a new process that has a copy of all the variable values in the main process. It's also fairly stable across different underlying operating systems, making it a good choice for this application.

Once the process has been created, the values change independent of the main process, so we need a way to communicate between processes. A common way to synchronize information between the main process and the worker processes is to create one or more shared queues, using the `Queue` class ❷. Typically, you want to create a queue for each direction of communication you want to support. In this case, the `p` queue will be used to send work objects to the worker processes, and the `q` queue will be used by the worker processes to send the solutions back to the main process.

Once we have our queues, we loop over each floor in the list and create a solver process. The `target` parameter tells the process what function to call when the `start` method is called from the main thread ❹. The target for these processes is `mp_agp_solver` function ❸, which we'll go over in a moment.

Notice that we pass in the two queues to every `mp_agp_solver` process as it is created. These work queues are how the different subprocesses will communicate with the main process. The main process will send work to the subprocesses using the `p` queue and receive the results back using the `q` queue. Once all the processes have been instantiated, we add each floor to the work queue by calling `p.put(w)` ❺ (where `w` represents the piece of work being sent—in this case, a `Polygon` representing the floor of the gallery).

At this point, the solvers can begin their work. As they complete the solution for each floor, they'll place the output into the return queue. The main thread continues to look for results in the queue until it receives a solution for every floor (based on the solution count matching the floor count). When it finds a result in the queue, it appends it to the `results` list ❻. Once all the results are received, the main process loops over all created processes and calls the `join` ❼ function, which essentially ends the process's execution.

Next, let's cover the changes necessary to the solution code to make it pull the work from the queue. Listing 12-8 shows the `mp_agp_floorplan` function that's called as the target of each subprocess.

```
def mp_agp_floorplan(p, q):
    work = None
    while work == None:
        floor = p.get()
    --snip--
    q.put(solution)
```

Listing 12-8: Modifying the solver for use in multiprocessing

NOTE *The full code for this function is in lines 47–77 of the* graph_shapes.py *file.*

The major change to the solution code from Chapter 11 is that we now add a `while` loop that tries to continuously get a floor to solve from the incoming queue `p` with `p.get`. Once the work object is received, the snipped portion of the code performs the steps for solving a floor plan we defined in the previous chapter, including converting the polygons to a graph, tessellating the floor shape, applying the greedy coloring algorithm, and so on. Once the floor plan has been solved, the graph object, Triangle result, and the coloring solution are packaged into a dictionary called `solution`, which is passed back to the main process using the outgoing queue with `q.put`.

Using parallelism to speed up your application and allow concurrent operations is an excellent way to move from the concept phase to a full-fledged application: most users have come to expect snappy performance. To get the most out of Python's parallel processing options, you should read the documentation for both the threading and the multiprocessing libraries. Thankfully, the developers of the multiprocessing library (the later of the two) had the good sense to model its API after the threading library, which was already very popular.

Adding a Graphical User Interface

Perhaps the largest single change from a proof of concept to a minimum viable product is the addition of a graphical user interface (GUI). Most users expect to be greeted by some visual workspace when running a program, and adding one makes your program accessible to the general public. Unfortunately, designing and developing a GUI can get very complex very quickly; there can be hundreds of components like buttons, text, and images that need to be managed. Users have different sizes and types of screens, which means laying out visual elements properly requires lots of additional code that you must account for. Furthermore, graphic elements like buttons aren't static. When you click a button, the program provides some kind of audio or visual feedback (such as darkening the button to make it look like it's been pressed). While this isn't a book on programming graphics in Python (a topic on which there are numerous tomes already), I couldn't totally ignore the topic: the graphics code makes up a large percentage of the project's total code base. Thankfully, PyGame has a collection of tools to help ease the pain. I'll cover some basic examples here to keep the code short and understandable.

Displaying and Managing Images in PyGame

The `Display` and `Surface` classes make up the base of the graphics platform. The `Display` class handles interfacing with the user's screen and contains parameters related to the video system as well as code to modify how the interface is displayed onscreen. The `Surface` class holds collections of elements that should be displayed together. Each copy of the `Surface` class is like a blank canvas. Typically, you'll have one `Display` class and one or more `Surface` classes to handle different (visually distinct) sections of the application (*https://www.pygame.org/docs*).

Listing 12-9 shows the simplest example of using the `Display` class.

```
  import pygame
❶ pygame.init()
❷ background = pygame.image.load("guggenheim.tif")
❸ screen = pygame.display.set_mode(background.get_rect().size, 0, 32)
  background = background.convert()
  running = True
❹ while running:
      for event in pygame.event.get():
          if event.type == pygame.QUIT:
              running = False
    ❺ screen.blit(background, (0, 0))
      mid_x = screen.get_width() / 2
      mid_y = screen.get_height() / 2
    ❻ pygame.draw.circle(screen, (0, 255, 0), (mid_x, mid_y), 50)
    ❼ pygame.display.flip()
❽ pygame.quit()
```

Listing 12-9: Displaying graphics using the PyGame display module

We start by initializing a PyGame application with `pygame.init` ❶. The `init` function performs a series of background steps that, among other things, let your system know Python needs to interface with the video display. We load the background image (in this case, the Guggenheim floor plan) using the `pygame.image.load` ❷ function. The `background` variable now holds a copy of the image data, which we can use to retrieve the width and height required to display the image with `background.get_rect().size`.

We pass this information as the first parameter to the `pygame.display.set_mode` function ❸ to create a `Surface` object that's the exact size of the background image. Before we can display the image, though, we need to convert it from the intermediate data format PyGame uses into a format that can be drawn to the screen faster. We begin the program loop ❹ by defining a sentinel named `running` that will remain `True` until the user executes the `pygame.QUIT` event. Everything inside the program loop is executed on each run, which allows us to update the graphic elements and create basic animations.

We tell PyGame to draw the background image on the next update by calling the `screen.blit` function ❺ with the converted background image object and the location on the `Surface` to place the image (based on the location of the image's upper-left corner). We can then place additional graphics on top of the background floor plan.

We first add a circle to the display using the `pygame.draw.circle` function ❻. The circle doesn't mean anything at the moment, but it does show how we can combine background images and shapes drawn with code. We pass the function the screen to draw on as the first argument. Next, we choose a color to draw with; you can use an RGB tuple (as shown in the example) or a hexadecimal color code. After that, we pass in the midpoint location (the center point for the circle) to tell PyGame where to place the drawing relative to the screen. Rather than hardcoding these values, we can calculate them relative to the size of the screen. We place the circle in the middle of the screen by taking half of the screen's width and height parameters as the x- and y-coordinates, respectively. The benefit of using relative position is that we don't need to change the location of objects manually if we resize the screen. The drawback is that the logic for laying out lots of graphics can become fairly long and tedious to develop.

Once we've laid out all the necessary graphic elements, we call the PyGame `display.flip` function ❼, which updates the whole screen area with any graphic changes. At this point, the whole loop then starts over. The loop continues until the user executes the `pygame.QUIT` event, which sets `running` to `False`; on the next loop check, the `while` loop is exited and the `pygame.quit` function is called ❽.

Figure 12-2 shows the result of running the code in Listing 12-9.

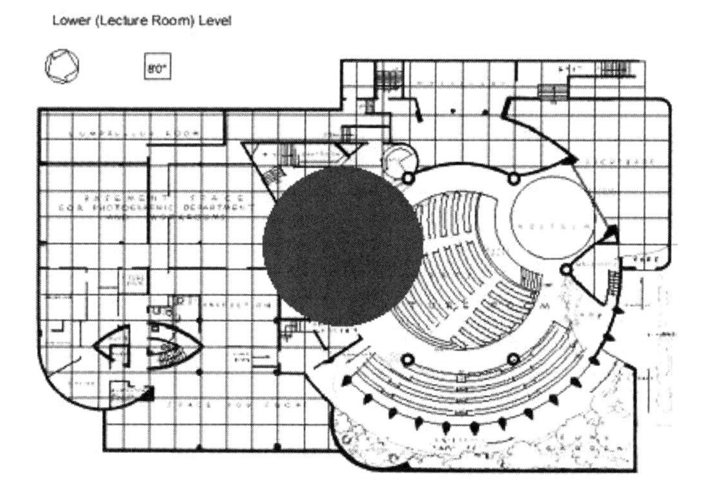

Figure 12-2: Drawing graphics onscreen with PyGame

This figure shows the loaded floor plan for the Guggenheim museum along with a large, black circle (which will be green when you see it onscreen) in the center of the image. The circle is on top of the floor plan because of the order in which we drew the items. Of course, drawing a circle in the middle of the screen isn't very useful, so we'll combine the positional information from MOUSE_CLICK events with the ability to draw shapes onscreen for more advanced functionality.

I highly recommend you dive into the Display and Surface class documents, as we've only scratched the surface here.

Organizing Graphics with Sprites and Layers

The problem with the previous method of drawing to the screen is that we're drawing all the graphics directly on the same surface, which means that if we want to remove the circle at some point later in the code, we'd need to clear the screen and then redraw the entire background. You can imagine how this would scale. If you have dozens of components onscreen, you'd need to redraw each of those as well. Drawing and redrawing the screen can become a computationally expensive task, which will make the program appear choppy.

A better method is to split up the graphics into individual components that can be displayed, removed, or changed without the need to modify the entire screen. To do so, we use a special class called a sprite, which you might be familiar with if you've ever worked with animation software. A *sprite* is a combination of the visualization of an element and the code to interact with it. For example, when a user draws a polygon representing a portion of a gallery (called a Room in the code), the Polygon is placed in a Sprite object that adds functions for showing or hiding the room on the display, as well as code to add obstacles (holes) to the room.

Listing 12-10 shows a simplified version of the Room sprite class.

```
❶ class Room(pygame.sprite.Sprite):
      color = (0,0,255) # Default to blue
      WHITE = (255,255,255)
      def __init__(self, vertices, screen_sz):
          super().__init__()
        ❷ self.vertex_list = vertices
        ❸ self.surface_size = screen_sz

      def init_surface(self):
          self.screen = pygame.Surface(self.surface_size)
        ❹ self.screen.fill(WHITE)
        ❺ self.screen.set_colorkey(WHITE)
        ❻ pygame.draw.polygon(
              self.screen,
              self.color,
              self.vertex_list,
              0 # Filled polygon
          )
    ❼ def clear_surface(self):
          self.screen.fill(WHITE)
          self.screen.set_colorkey(WHITE)
```

Listing 12-10: Creating a custom sprite class for Room polygons

NOTE *The* Room *class, as well as all the other custom classes, is in the* primitives.py *file.*

Every custom sprite class you write begins by extending the pygame.sprite.Sprite class, either directly, as shown here ❶, or by extending another class derived from the original Sprite class. The rest of the class definition is identical to other classes. You can assign class attributes and use the __init__ function to customize each instance of the class. In the __init__ function, we set the instance's vertex_list attribute (the exterior points of the polygon to be drawn) ❷ and the surface_size attribute ❸ (the size in pixels of the surface used to display the drawn polygon).

The init_surface function creates the actual Surface object in the attribute named screen, which will hold the drawn polygon data. By default, drawing a polygon on a screen will result in a solid background around the exterior of the polygon shape. When laid over the background image, this additional color will block portions of the floor plan from view, which is no good. We can make the background transparent by filling the surface with a color (white in this example) ❹ and then calling the set_colorkey function ❺ with the same color. The set_colorkey function makes the pixels that match the key color transparent. The effect in our application is to hide everything but the shape of the polygon so you can lay it over the floor plan without needlessly blocking sections.

NOTE *This is the same process used in* chroma key filming, *also called* green-screen filming. *Scenes are filmed against a bright solid-colored background, often green (hence the nickname), which is later made transparent so a false background can be inserted.*[1]

We draw the polygon to the surface using the pygame.draw.polygon function ❻. Note that the last parameter you pass in represents the line thickness to draw with. Setting the thickness to 0 tells PyGame to completely fill the polygon with the color you pass in. Make sure the color you pass to the set_colorkey function is different than the one used to draw the polygon, or you'll remove the polygon as well.

Finally, the clear_surface function ❼ uses the same set_colorkey trick to make the surface completely transparent, which is useful for temporarily hiding the polygon on the screen, without completely removing it.

The Room class also has several functions that it inherits from the parent Sprite class, such as add, remove, and update, all of which allow you to control the sprite's behavior after it has been created. Listing 12-11 shows how you can use the Room class to define a polygon overlay.

```
gallery_poly = [(20, 10), (20, 20), (55, 148), (145, 145)]
--snip--
poly_sprite = Room(gallery_poly, background.get_rect().size)
poly_sprite.init_surface()
screen.blit(poly_sprite.screen, (5,5))
--snip--
```

Listing 12-11: Using the Room *sprite to display a polygon*

NOTE *The full code for adding a* `Room` *to the project is in lines 306–336 of the* state_manager.py *file.*

First, we define a polygon to display as a list of vertices named `gallery_poly`. You can take these points from user input (like a series of mouse clicks) or load them from a file, as long as the data is in (*x*, *y*) format.

We define the sprite to hold the polygon data by calling the `Room` initialization method defined in Listing 12-10. We prepare the polygon for display by calling the `init_surface` function. Finally, we call the `blit` function on the main display `screen` and pass in the polygon's screen attribute with `poly_sprite.screen`. Blitting the sprite's screen onto the main screen tells PyGame to update the display with the information from the sprite.

You can see the result of this code in Figure 12-3.

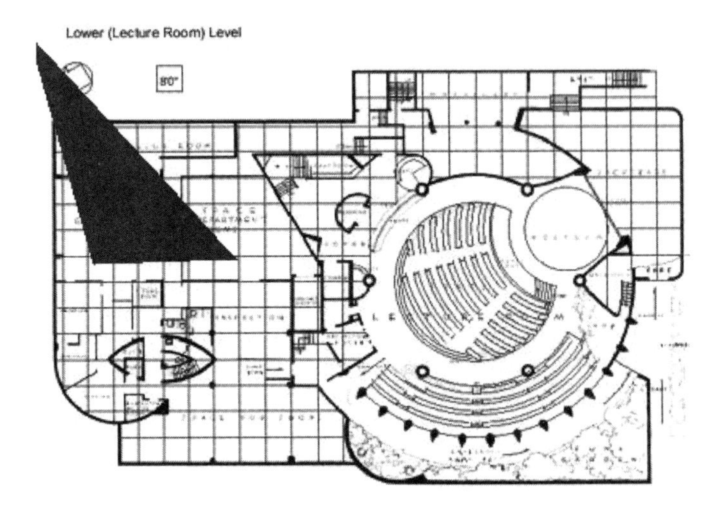

Figure 12-3: Drawing a polygon with a sprite

By combining the techniques of capturing user input, drawing directly to the screen, and using sprites to manage more complex graphic elements, you can come up with exquisitely detailed interactive displays to allow your users to complete their tasks. My advice for interface development is to start simply and build up, rather than trying to develop the whole UI at once. For example, start with the ability to draw a polygon using mouse clicks and keyboard shortcuts before adding graphic buttons and menus. When you're ready to develop a more visually appealing UI (such as one with buttons and checkboxes for configuration options), check out another library named Phil's Game Utilities (PGU), which is written to complement PyGame's display classes with a large number of predefined visual components (like the aforementioned checkboxes). Using PGU, you can quickly develop context-aware menu options, animated buttons, and other polished visual elements users have come to expect. The downside is that it takes a lot of code to make the magic work. Showing and explaining the code for any nontrivial GUI would take more pages than I have to cover the rest of the art gallery project.

Saving and Reloading Project Data

We've discussed the user's desire to work on complex projects over a number of working sessions. The desire to save and load work is almost universal when it comes to practical software. Python offers developers a wide variety of options for saving and reloading data. You have traditional methods, like databases and flat files, along with more modern options, such as pickle, a library for saving and loading Python objects.

The exact method for saving and loading data has to be designed with your specific architecture in mind. For example, if you're developing your application to run in the cloud, you might want to avoid saving data to flat files, or even local file-based databases like SQLite. At the very least, you'll need to plan for storing the data in a manner that can be accessed from your cloud infrastructure (more on this in the next chapter).

Saving to a Dictionary

I chose to save the data as several files, grouped together inside a compressed (that is, zipped) file. The majority of the data will be stored in JSON files, which can easily be written from most dictionaries. To make life simple, I added a function called `to_dict` to each custom class, which returns all the data necessary to redefine the object from an empty copy of the class.

Listing 12-12 shows an example of the `to_dict` function for the `Room` class.

```
def to_dict(self):
  ❶ return {
        "vertex_list": self.vertex_list,
        "name": self.name,
        "color": self.color,
        "floor": self.floor,
        "surface_size": self.surface_size,
      ❷ "obstacles": [o.to_dict() for o in self.obstacles]
     }
```

Listing 12-12: Creating a dictionary representation for the `Room` *class*

This function returns a dictionary object ❶ that is designed to be JSON serializable. You can include nested objects that also have a `to_dict` function. The `obstacles` key ❷ will hold a list of dictionaries, each of which defines an `Obstacle` object to include in the room during reload.

Converting a dictionary to a JSON string is simple using the json library (part of the standard libraries), as shown in Listing 12-13.

```
room = Room([(2,2),(2,4),(4,4)], (10,10))
--snip--
room_str = json.dumps(room.to_dict())
with open("/tmp/project/floors.json", "w") as f:
    f.write(room_str)
```

Listing 12-13: Saving a JSON representation of an object

We start by creating one or more objects. The call to the `Room` initialization creates a room shaped like a triangle with a display area of 10×10 pixels. We create a JSON representation of the room by calling `json.dumps` and passing in the dictionary representation of the `room` object. The `json.dumps` function returns a string that can be used to safely store or transport the data across a network.

After creating a JSON string for every object we want to save, we write the strings into a file called *floors.json* inside a temporary directory. If you're on a Windows machine, you'll want to change the references to the *temp* directory to something like *C:\Users\Temp* depending on your version. At this point, you could incorporate the `DataSaver` from Listing 11-8 to save the `Triangle` files (if the user has created any tessellations), but since we already covered that code, I won't show it here.

Once all of the data objects and resources are saved as files in the temporary directory, we create a compressed archive file for easier storage. We can do this with the `make_archive` function from the `shutils` library like so:

```
shutil.make_archive("output_file.agp", "zip", "/temp/dir")
```

The first argument is a string that will be used to name the output archive. The second argument is the format to use when compressing the archive. Several options are supported, but the most universal, by far, is the *.zip* file format. The final argument is the path to the directory that holds the temporary files. At this point, you'll have a compressed file called *output_file.agp*, which can be examined with any common unzipping tool (I personally like 7Zip because it's free and cross-platform).

Next, we'll look at reloading the data from the JSON strings.

Loading from JSON Files

Reloading files works in the reverse order as saving. The program takes in a compressed archive, unpacks it to a temporary directory, and attempts to rebuild all of the objects contained within. Unpacking an archive to a directory is fairly straightforward, as shown in Listing 12-14.

```
from zipfile import ZipFile
import os
os.mkdir("/tmp/project")
with ZipFile("output_file.agp", "r") as zipf:
    os.cwd("/tmp/project")
    zipf.extractall()
```

Listing 12-14: Extracting files from a previously created ZIP file

We start by creating the temporary directory where the project files will live with `os.makedir`. We then open the archive by calling the `ZipFile` class with the path to the archive and the mode to use. Using the `with` construct allows us to keep the file data in memory as the variable `zipf` until the indented block of code is complete. Once the code block finishes, Python will automatically close the file.

We change the working directory into the temporary project directory with `os.cwd`. Finally, we extract the files to the temporary directory by calling `zipf.extractall`.

Once the archive is extracted, we can load the data back into a dictionary with the following code:

```
with open("/tmp/project/floors.json") as f:
    room_dict = json.loads(f.read())
```

Essentially, this code block opens the *floors.json* file that was extracted in Listing 12-14 and then passes the contents of the file to the *json.loads* function. The result is a dictionary object that has the same structure and data as if you had called the `to_dict` function.

The final step is to convert this data back into the proper object classes. To aid in the process, I recommend adding another function called `from_dict` to each class as you develop it. The `from_dict` function is the complement of the `to_dict` function in that it converts a dictionary-like object into object parameters.

Listing 12-15 shows the `from_dict` function for the `Room` class.

```
def from_dict(self, p_dict):
    for k in list(p_dict.keys()):
      ❶ if k == "obstacles":
            for k2 in p_dict[k]:
              ❷ obs = Obstacle([])
              ❸ obs.from_dict(k2)
                obs.init_surface()
              ❹ self.obstacles.append(obs)
        else:
          ❺ setattr(self, k, p_dict[k])
```

Listing 12-15: Reloading a `Room` from a dictionary

To simplify the process, there are only two cases of interest for the `Room` class: the case where we're dealing with the list of obstacles and the case where we're dealing with all other attributes. If the code is parsing the `obstacles` attribute ❶, we loop over all the obstacles in the data. For each, we create an empty `Obstacle` object ❷ and call the `from_dict` function ❸ on the data to recreate the object with all the appropriate attributes. Calling `init_surface` prepares the `Obstacle` object for display when the time comes. Finally, we append it to the room's `obstacles` attribute ❹.

In the event where the parameter being processed isn't an obstacle, we simply add it as an attribute of the object being created using the `setattr` function ❺. Using the `setattr` function allows us to expand the definition of each class (for example, adding a new attribute to the `to_dict` function) without needing to update the `from_dict` function as well. Because we can update one function without impacting the other, we call these two functions *loosely coupled*. Loose coupling is a good goal to strive for in a production application because it reduces the amount of effort needed for ongoing development. We can use the from_dict function to recreate the room object from the file data with the following code:

```
room = Room([], (1,1))
room.from_dict(room_dict)
```

First, we create an empty Room object to hold the attributes; then, we call the from_dict function and pass in the dictionary room_dict. At this point the code is back to the same state it was in at the beginning of Listing 12-13.

NOTE *Both the* to_dict *and* from_dict *functions are inside the* Room *class in the file* primitives.py.

Running the Example Application

I've included a working example of the application in the chapter's supplemental materials. It includes the features we've discussed up to this point. You can use the application to play with solving your own floor plans, then dive into the code and start improving it! Once you've navigated to the project folder, you can see the application's help screen with this command:

```
$ python poly_draw.py -h
```

The output will be a list of options like these:

```
--snip--
Usage: poly_draw.py [options]

Options:
  -h, --help            show this help message and exit
  -r RESUME_FILE, --resume=RESUME_FILE
                        Load Room shapes from a previously saved file
  -b BGD_FILE, --background=BGD_FILE
                        Background PNG to load (if not resuming)
  -o OUT_FILE, --out-file=OUT_FILE
                        File to save shapes to
  -e EPSILON, --epsilon=EPSILON
                        Sets near-miss distance
  -f, --use-feet        Convert answers from meters to feet
```

We've already covered the first option, -h, which prints out this helpful menu. The -r option allows you to resume sessions by passing in the location of a compressed project file created using the save feature we discussed earlier. The -b option allows you to set the base floor plan's background image when you're starting a new project. The -o option allows you to specify the filename to save shape data to when the save command is called (with the S key), and the -e option allows you to change the value the program will use for epsilon. If you have trouble clicking vertices, you can try increasing this number a few pixels at a time until you achieve the desired result. Turning it up too high can cause unexpected behaviors, though. Finally, the `-f` option tells the application to use feet as the unit instead of the default meters when scaling the solution.

You can try the program out with a floor plan like the example of the Guggenheim as follows:

```
$ python poly_draw.py -b guggenheim.tif -o myfloorplan
```

Replace the reference to *guggenheim.tif* with the name of your file. The application will ask you to enter a name for the floor plan in the console. For now let's call the floor "main." After you press ENTER, the application will open the Guggenheim image and enter the scaling state. Click two places in the image to draw a line. The console will then ask you to input a length for this line. This is how the application will figure out the real-word distance of the scaled image. You can now start drawing the rooms on the screen by clicking around with the left mouse button. When you want to finish the room, connect the final point back to the initial starting point to create the closed shape of the room. When you click the starting point to close the room, the console will ask you to input a name for the room. Enter something like "Gallery" and press ENTER. You should see the inside of the area you traced out change to green. Finally, try saving your work with CTRL-S. You should see the console spit out a message like this:

```
Saving...
Saving 1 floors.
Saving 1 rooms.
Successfully Saved: myfloorplan
```

If you list the files in the directory, you should now see a compressed file named *myfloorplan* that contains the floor and room data you just created. You can now safely exit the program using the X button at the top of the window. You can reload your previous save and begin work again with the following command:

```
python poly_draw.py -r myfloorplan -o myfloorplan
```

It's actually best practice to change the `-o` filename by appending a version number between edits. This way, you can always go back to a previously working copy if something goes awry.

There are many more functions in the example application and better ways to handle keypresses. There are also some incomplete features I've left for you to finish for practice. Read the code and play with the application to see if you can complete them.

Summary

We've covered a great deal of material in this chapter, but we've barely scratched the surface of user interfaces in Python and the multitude of events available in PyGame. You've seen how you can use these events to capture user input via the keyboard and mouse, but there are limitless ways you can expand on the concepts shown here to make your application more intuitive or enable power users (users that have a high level of familiarity and understanding of an application) to speed up their work. As mentioned, for a more visual interface, I encourage you to look into Phil's Game Utilities, especially if you want your application to be accessible to the largest number of users.

A polished visual display is key to gaining interest from less technical users. There's an entire field of study devoted to understanding how people interact with systems called *human–computer interaction (HCI)* research (*https://en.wikipedia.org/wiki/Human–computer_interaction*). As a developer, you can use HCI research to quantify how users interact with your application and locate areas for improvement. As a security analyst, you can use knowledge of HCI to plan interface controls that make managing privacy and security more intuitive. One particularly good source on the topic is the book *Research Methods in Human–Computer Interaction* (Wiley, 2014).[2]

We also covered one possible way to save data between user sessions using JSON files. Saving the data in a human-readable format like JSON allows you to troubleshoot the save and load functions easily during development. We'll discuss other potential options for saving the data in the next chapter, where we'll also cover different options for distributing your application to the masses, including moving the application into the cloud.

13

DELIVERING PYTHON APPLICATIONS

Once you feel you've met the requirements for a minimum viable product, it's time to focus on the delivery pipeline. A *delivery pipeline* defines how your users will get your application and any future updates. Truthfully, I tend to start my projects by defining these attributes, as they can make some development choices more or less advantageous. For example, if you decide to deploy your application to the cloud, saving the data as local files doesn't make as much sense as it would if you plan to deliver your code as a local package. In this chapter, we'll take a high-level look at four potential delivery pipelines. Each method has a deep pool of resource material to help you deliver your project, so I'll focus on the important considerations, strengths, and weaknesses of each method.

In addition to discussing the delivery aspects of each method, I'll talk about the removal process. One part of an MVP that people often neglect is the uninstall capability. Having a good, clean uninstall function for your application is, in my opinion, one of the keys to being a good software vendor. The objective for your uninstaller should be to leave nothing behind for the user to clean up. As you'll see, some methods make this easier than others.

One major influence on which delivery method to choose is whether or not you plan to monetize access to the application, and how. For example, if you want to charge users a subscription fee, you probably want to skip to the sections "Distributing with Cloud Microservices" or "Licensing with PyArmor," both options that would allow you to define who has access to your application. The downside is that neither option is free, so if you don't plan to charge for access to your application, it may be more cost-effective to distribute the application via GitHub using the setup script or prepackage the application with all the files needed to run the application.

Using Setup Scripts

The simplest choice for distributing a Python application is to use a special configuration script named *setup.py*, which configures the underlying system with the libraries and supporting files necessary to run the code. You've probably run into this method if you've installed a Python module after manually downloading a GitHub repository. More generally speaking, installing code packaged using this method requires your users to install a library called setuptools, which handles the installation based on the structure you define in the setup script. You can learn more about the structure and options for the setup scripts by reading the PyPi documentation (*https://pythonhosted.org/an_example_pypi_project/setuptools.html*).

The major benefit to using this method is that you can get your project hosted on PyPi, which will then allow your users to install it simply using the pip tool. When you install a project from PyPi, the pip tool pulls the appropriate version of your project from a storage location you define (most often, a public GitHub repository, although there are other options as well). Your users won't need to manually download repositories or run setup scripts, all of which is handled in the background.

However, there are some definite drawbacks to this method. First, it's very difficult to monetize the code. There are no controls to stop a user from copying the source code to another machine. There's also no native way to remove the code once it's installed, which means even if you found a way to monetize access, you'd be limited to charging a one-time fee for lifetime access. There would be no way to enforce something like a subscription plan. Second, the setup script relies on your users installing the application from the command line. This is fine if you expect your users to be familiar with that process, but this isn't the best choice for delivering applications to the general population. Finally, installing an application this way makes changes to the user's underlying system. It installs the packages and configures them. While this works most of the time, making any changes to a system runs the risk of corrupting something, colliding with existing files, and so on. If the user doesn't install your application in an isolated virtual environment, there's the very real risk of running into incompatible library versions with some other application the user installed.

Removing applications installed with this method poses its own problems and usually requires the user to remove the requirements and other resources from their system. If you opt to go this route, I recommend pushing your users to use separate virtual environments. If the user hasn't set up the application in an isolated virtual environment, trying to helpfully uninstall a dependency may break other applications on their system. On the other hand, if your users do install everything in its own virtual environment, uninstalling is as easy as deleting the environment.

I recommend using the setup script method for open source applications and smaller projects that aren't backed by the resources for a more complex delivery pipeline. You can deploy a module and setup script to PyPi in a matter of minutes once you have an account. Setup scripts are also a good entry point for understanding more complex deployments, like cloud services, because at some level, all these methods need a way to understand what dependencies must be present for the code to function. Overall, this is a solid delivery method every Python developer should be familiar with.

Packaging with Python Interpreters

The next option aims to address some of the shortcomings of the setup script method by taking some work off the user. The idea is to package the code, support files, and the Python interpreter to run it into a single archive that can be delivered to the user. The user then simply needs to extract the archive to a directory on their system and they're ready to run the application.

This allows for better monetization than the setup script. By hosting the packaged application download behind a website, you can charge users for each new version or you can charge a monthly subscription fee to the site that includes access to the latest versions for download. There's still nothing to stop a user from paying once and keeping that version forever, but they also have incentive to maintain their account for access to the latest features.

To handle the packaging, I use PyInstaller, a free application to help collect the necessary files to make your program stand-alone, meaning able to run without configuring the underlying system. Packaging with this method is often called *freezing* an application because it collects a copy of the current version of all dependencies and the Python interpreter installed on the system, and then it packages them such that the included interpreter will use only those packaged libraries to operate. The advantage here is that you don't need to worry about what version of a package is installed or whether it will conflict with other applications on the user's machine. The disadvantage is that if you need to update one of the underlying libraries after freezing the application—for instance, to mitigate a security risk in one of the dependencies—it requires releasing a patch or a new build of the application for distribution. If a user doesn't apply the patch or download the latest version (all too common), their system is left at risk.

The other drawback is the size of most frozen applications. To make sure the internal code functions properly, the entire standard library is usually frozen along with the other dependencies. The large code base and the Python interpreter mean that even simple applications may end up being multiple megabytes. PyInstaller does what it can to minimize the bloat, and you can configure it to reduce the weight even further, but ultimately there will always be additional bloat with this method.

Like energy, complexity doesn't just disappear. Taking the complexity off the user means putting it onto yourself. For the frozen delivery method to work, you'll need to create a different package for each type of system you want to support. For example, you may end up with one package named *agp_linux64_amd.tar.gz* intended for users on a 64-bit Linux system that has an AMD, one named *agp_win64_intel.zip* intended for users who have 64-bit Windows running on an Intel platform, and so on. To package each of these, you need access to a copy of the underlying OS that can be used to package the system files. To run the application during development, I use VirtualBox with a copy of each OS as a virtual machine preconfigured with the proper dependencies and version of Python. I like this method because it allows me to automate the build process for several platforms at once using the VirtualBox-manager application and some custom scripting (*https://www.virtualbox.org/wiki/Documentation*).

For Windows, you face a unique circumstance. At the time of writing, some of the necessary drivers are protected by Microsoft licenses. Redistributing these libraries within your application without paying a licensing fee could be a breach of the Microsoft terms of service, and may even result in you being held liable for any perceived loss of revenue. The caveat is that if the end user already has a copy of the libraries (as is usually the case), then sending them the application wouldn't violate the Microsoft agreement. Suffice to say: when and how you can deliver a prepackaged Python application for Windows is a bit of a gray area. Don't interpret this as legal advice; I am not a lawyer. I insist you consult with an attorney in your area who specializes in intellectual property disputes around technology licenses. They'll be able to help you steer clear of any legal risks.

Freezing your application might be a good choice if you want to deliver a stand-alone version of it for users. The benefit of a stand-alone project is that it's easy to set up and remove from a machine—uninstalling can be as simple as deleting a folder. In a lot of cases, frozen applications can even be run from a USB storage device, meaning you can carry it with you anywhere, and you won't need to install the code on a system to use it!

Distributing with Cloud Microservices

Deploying to the cloud means different things for different people. You could argue that hosting the packages from the previous messages using some data storage service (like Amazon's S3 or Google's Cloud Storage) and hosting the website in a virtual machine served by the same vendor constitutes a cloud deployment. It's true that the delivery pipeline is a cloud service at that point, but the application itself would still be downloaded and run locally by the user, so I don't consider it a true cloud service.

To me, a cloud deployment runs the majority of its functional code from infrastructure hosted by a service provider (like Google or Amazon), meaning your application is served to users in a functional state, rather than sending them source code to run. For the rest of the chapter, I'll refrain from speaking about any one service provider. The two major global cloud providers, Google Cloud Platform (GCP) and Amazon Web Services (AWS), maintain approximately equivalent features, so I think it's more beneficial to discuss the concepts generally. You can take these concepts and learn how to apply them using your particular provider.

Rather than hosting code to download, the user will generally be given access to your software via a website of some sort. It's possible to have a user-side application that acts as the interface to the cloud structure, but this is less common because it adds complexity to an already complex process. The code is broken up into small chunks called *microservices*, each of which handles one small part of the application.

Figure 13-1 shows a simplified microservice architecture for the AGP project.

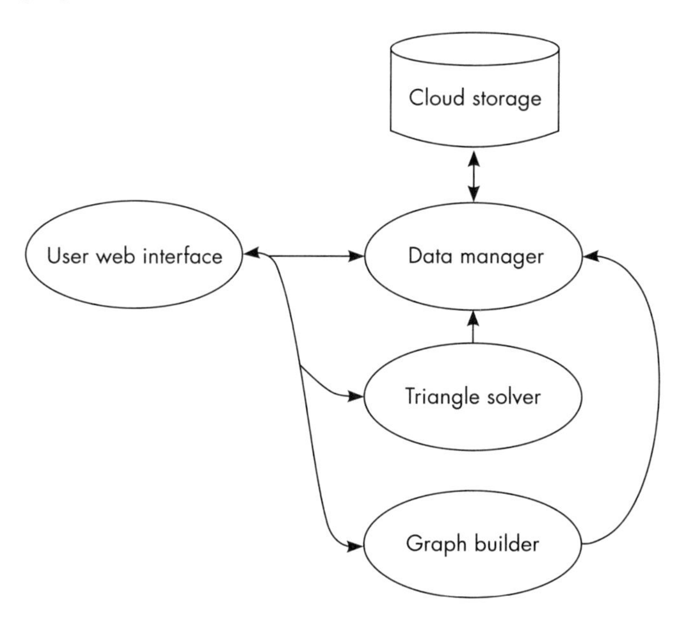

Figure 13-1: Microservice architecture diagram

Each oval represents a small part of the application run inside a virtual machine with just enough resources to execute the function and then disappear once it's no longer needed. The key to a good microservice deployment lies in cleanly separating the functions into services and efficiently managing the communication between the services (the black arrows in the figure).

In Figure 13-1, I've divided the project into four services. The user interface is moved into a website that most likely uses HTML5 for the interactive drawing and JavaScript for communicating with the rest of the services. JSON is a good choice for the communication protocol since the communication between services is mostly handled using HTTP requests, and both languages involved (Python and JavaScript) handle the format easily. The Data Manager service holds the functions for saving and loading project data for users.

One limitation of a lot of microservice designs is that they lack a permanent filesystem to serve files from. You can overcome this by creating a persistent storage location, or you could make the storage location a persistent database instance. In fact, with a bit of ingenuity, you can make any network-accessible storage location work for this purpose. In any case, placing the Data Manager service between the other services and the storage container means the Data Manager is the only service that needs to know how to read and write from the Cloud Storage container. If you decide to migrate to a different method of storage later, you'll only need to update one service.

The Graph builder service contains all of the functions for the application to manage the graph representation of a gallery. It communicates with both the User Web Interface (to take in the JSON data representing the graph) and the Data Manager (to save the information once completed). The Triangle Solver service contains the functions for managing the polygon representation of a gallery, including the code to ultimately solve each floor using the Triangle library. It also talks with the User Web Interface service and the Data Manager service to handle the input and output for the code.

Docker is great for microservices because it allows you to configure each virtual machine to contain only the pieces necessary to run the service's code, making them faster to create and more secure to operate. You can use Docker containers to define these tiny machines on most cloud providers (*https://docs.docker.com*). Furthermore, you can use a container orchestration platform like Kubernetes to automatically manage the creation and deletion of each service container as needed.

Automatically creating more instances of your application to serve to users is called *horizontal scaling*. Taking advantage of the horizontal scaling capabilities of your platform will allow your application to seamlessly adapt to changes in processing needs. You can define rules for each service individually, meaning you'll scale only the parts of the application that need to and leave the rest alone. For example, suppose your application has 20 users who simultaneously request the solution for different floor plans. With a traditional architecture, the Triangle service would have to handle all 20 requests, so the last user in the queue will have a longer wait than the first. With horizontal scaling, the orchestration engine will see the increase in demand and add 19 more copies of the Triangle Solver service. The additional copies all run in parallel, so all the requests can be handled simultaneously. On the other hand, 20 users on a web server is usually fine, so you wouldn't want the orchestration platform to add more copies of the

User Interface Service. By configuring automated horizontal scaling rules for each service individually, you can save yourself hours of maintenance down the road. This is the third form of parallelism you can take advantage of: *hardware parallelism*. It's similar to process parallelism, discussed in Chapter 12, but the work is spread across different machines instead of different cores on the same machine.

The cloud microservices method is probably the most complex to achieve initially, but the benefits are also numerous. We've already seen how its flexibility can allow for quick iterations and reduce maintenance time. Another benefit is that you can easily monetize the application with much greater control. Since the source code is never sent to the end user, they must maintain their account to continue accessing the service. There's usually little to nothing for the user to uninstall if they choose to stop using the service as well. All a user needs to do is terminate their account and the service is gone, making it the cleanest exit method from this perspective.

You can run multiple versions of the application to serve users at your discretion. Most cloud service providers offer application layer traffic routing, which allows you to intelligently direct traffic to different copies of your application based on some rules you define. Kubernetes also has some traffic routing capabilities that can be used to achieve the same effect. You can use the traffic routing ability to selectively beta-test changes before distributing them to all users or to define separate testing and production versions (called *A/B* or *blue/green testing*).[1]

Finally, updates are entirely under your control. Users automatically have access to the newest production version as soon as you update it. Typically, microservices are built in several stages. The stages can vary from project to project but roughly follow these phases: code push, continuous integration testing, Docker container build, and finally, service deployment. The code push is probably already familiar to you. It occurs when you push some changes to the code repositories (for example, with the command `git push`). Pushing the code triggers the continuous integration tests. These tests are designed to ensure that you don't introduce any common bugs with your changes (*https://circleci.com/blog/proactive-integration-testing*).

The major drawback is the monetary and time costs associated with building and maintaining the network of services required to make the application behave seamlessly for users. Each cloud provider comes with its own unique way of implementing the pieces, and the pieces themselves require you to understand auxiliary applications like Docker and Kubernetes to work. If you decide to deploy your application in the cloud, you'll need to spend some time learning the idiosyncrasies of the platform you choose. It also doesn't hurt to have a development team to support your cloud deployment efforts.

You can't expect to become an expert in all the different components necessary for a solid cloud deployment pipeline. I've been lucky through my career to have the pleasure of working with some of the best cloud engineers in the field, and what I learned was the value of a strong team made up of different specialists. Having someone who can focus on the architecture while another person focuses on the user interface while a third writes the main service code means the work can get done much quicker; this is parallel development. Having a team also allows each person to maintain focus on the areas where they are most knowledgeable, and your project to benefit from the added expertise. Of course, running a development team brings its own host of problems: conflicting personalities, failures on delivery dates, and so on. Deciding to use a development team means you also need someone who will be responsible for communication and coordination among the team members[2] (called the *project manager*). Cloud deployments are at the heart of all modern software-as-a-service (SaaS) companies because the long-term benefits far outweigh the initial development cost. Of course, cloud deployments may not be the best choice for a small project with only a few planned users.

Licensing with PyArmor

The next method is a bit of an oddball. PyArmor is a command line tool for obfuscating Python source code. It's conceptually similar to the standalone application in the sense that you still deliver an executable to your users, but PyArmor tries to ensure that your Python applications are saved and run only from approved machines in an attempt to protect your intellectual property and help monetize executables you deliver to your users. *Obfuscation* is the process of hiding the structure and operation of the code to render the application inoperable without the proper de-obfuscation technique. An example of obfuscation might be changing the string `"Hello from PyArmor"` to something like `"H7ejl8l3ocb1fRr4osm9blPjy9Afr4mvoOrp"`. The application obfuscates constants and literal strings as well as the runtime code of each function. If someone attempts to read your source code, either at rest or in memory, they'll be met by a wall of gibberish.

Obfuscating your application's source code also enables you to bind your application to a single machine and expire the application remotely. Your code is hidden behind a startup application that contains the proper de-obfuscation technique. PyArmor startup scripts check for a license file that was created when the user installed your application. The license file uses some unique machine attributes to ensure it's run on the same machine at each start.

You can also define a license server that manages the validity of each license. On each start, your application calls out to the license server with its license identifier. Your server can then respond with an authenticated message saying whether or not the application should allow itself to execute. Of course, this relies on your user having network access every time they want to run your software, which may or may not make sense for your project.

Obfuscation shouldn't be confused with encryption, and it certainly doesn't offer the security of a strong encryption scheme. With encryption, you have some mathematical proof of security using some form of secret information. The bigger and harder to guess that secret is, the more secure the application. Obfuscation, on the other hand, is more like camouflage: it's meant to hide the code from casual attempts at reverse engineering or bypassing the license restrictions. But once someone understands how the code has been obfuscated, they can always reverse the process. Going back to my previous example, if you examined the second string more closely, you may have realized that all I did was insert random characters between the letters of the phrase. Next, I replaced the spaces with a constant character, `b`. By reversing this process, replacing the even-numbered `b` characters with spaces and stripping every other character out, you can convert the jumbled string back to the original phrase. As if that weren't bad enough, the part of the startup script that handles the de-obfuscation can't be obfuscated itself, meaning the technique used is available to anyone who wishes to look for it. (If it could be obfuscated, you'd create a feedback loop as you'd need a de-obfuscator for the de-obfuscator.) To borrow a phrase used in the lock-picking world: "It's good enough to keep honest people honest." Depending on the level of obfuscation, this may only hinder a talented reverse engineer by a couple hours.

Although there are some obvious limitations, I wouldn't completely dismiss PyArmor either. PyArmor may make a good addition if you plan to monetize the stand-alone package delivery method because it does add some control and monitoring to the process. You can't guarantee your controls won't be bypassed, but it certainly makes it less likely than a stand-alone application without obfuscation and licensing applied. Another potential use for PyArmor's licensing is tracking the number of active users. Even if you don't plan to monetize your project, having a license in place allows you to approximate the number of users by looking at which licenses have checked in (meaning the application was started). As your project gains popularity, you can use these estimates to gain investor interest or potentially sell your project to a larger SaaS provider.

Open Source Delivery

There's no better place to finish than with the option of open source delivery. By far the easiest method to deliver your project and to give back to the community is to openly license the source code for your project to everyone. By design, open source software licenses promote collaboration and sharing as they permit other people to make modifications to source code and incorporate those changes into their own projects. By hosting your code base on a public GitHub repository (or similar), you get the benefits of crowd-sourcing development, getting feedback from potential users, reducing hosting costs, and more. Open source projects promote collaboration from diverse perspectives. Different people from around the globe can come together and contribute. People have different and unique ways of solving problems, so including contributions from people with diverse backgrounds can push your project to a level of functionality you never would have achieved on your own, even with a traditional development team.

One common misconception is that "open source" automatically means you can't monetize your application. This isn't true at all! While it's true that most open source projects aren't started for money, maintaining a large open source project—Kubernetes, for instance—is a lot of work! It takes several full-time developers who probably want to get paid for their time, so open source projects will often spawn successful companies. Projects are often open-sourced in conjunction with a cloud delivery option to provide both a free and paid version. Companies will pay for these cloud-hosted versions to reduce the number of systems they have to maintain internally. Red Hat, the maintainer of one of the most popular Linux distributions for enterprise use, is one example of a large open source company that follows this model. While Red Hat continues to offer the open source version of many of its applications, it also offers paid customizations and remote support to maintain the business. In short, choosing to open-source your code often will reduce your stress and encourage a better result for your project, but you don't have to sacrifice monetization. I highly recommend you research the open source route when considering delivery methods for your application.

Summary

Deploying your application for general use can seem like a whole project in itself. As you've seen, there are several factors that should influence your choice of method. These include the number of users you plan to serve and whether you want to charge for access. As I said at the start of the chapter, you should start with the basic concept of the delivery platform in place when you begin your project. Once you've decided how to deliver your application, you can let the choice inform the rest of your development decisions, such as what storage options are available to your code.

At this point, you should have an idea of the available options and the positives and negatives of each. You can take these basics and learn more about the process that suits your needs best. Whatever method you choose, it's important to remember to think from a user's perspective as well as from a developer's. Be kind to users and give them intuitive ways to install, manage, and remove your application.

The internet is filled with resources for learning about software deployment plans, ranging from the very simple to the exceedingly complex. There are also excellent books covering various deployment technologies like Docker and Kubernetes (*https://bookauthority.org/books/new-continuous-delivery-books*). I recommend starting small and working your way up. If you've never used Git, jumping into a cloud deployment right away is going to lead to frustration. Start with something like PyPi, which will allow you to hone your repository management skills. Once you're comfortable with each of the underlying pieces, you'll be better equipped to tackle the large cloud deployment process.

And with that, we've come to the end of the book! If you've tackled all the concepts and projects, I congratulate you! I hope you feel that you understand the role that applied mathematics can play in your security tools. If you take anything away from this book, I hope it's the idea that you can tackle seemingly complex research topics with only a basic knowledge of math and an understanding of programming. Topics such as facial recognition, privacy monitoring, and social network analysis may be getting all the headlines at the moment, but the number of open research problems in the broader security field is huge, and they would all benefit from a talented and dedicated researcher like you. If there's a particular field that interests you, I encourage you to take the concepts you've learned and apply them to that field as well. The fields covered in the book all lend themselves incredibly well to multiple areas of interest, and when you mix and match them, you can achieve some very powerful analysis tools.

The scariest part of applying security to the real world, where a mistake could cost lives, is the need to make decisions in the face of uncertainty. Analysis tools like the ones presented in the previous chapters allow us to examine the world in different ways and to make the most informed decisions possible. You may not be able to remove uncertainty completely, but you can minimize its impact on yourself and those around you. Remember: security isn't just a job or career path, but a way of understanding the world. The future of security applications lies in the accurate collection, interpretation, and response to data collected from our physical and digital environments to aid us in that understanding.

NOTES

Chapter 3

1. Chensi Zhang et al., "Graph Theory Based Cooperative Transmission for Physical-Layer Security in 5G Large-Scale Wireless Relay Networks," *IEEE Access* 5 (2017): 21640–21649.

2. Peter D. Zegzhda, Dmitry P. Zegzhda, and Alexey V. Nikolskiy, "Using Graph Theory for Cloud System Security Modeling," in *Proceedings of the 6th International Conference on Mathematical Methods, Models, and Architectures for Computer Network Security*, eds. Igor Kotenko and Victor Skorminpp (Berlin: Springer, 2012), 309–318.

3. Linton C. Freeman, "Centrality in Social Networks Conceptual Clarification," *Social Networks* 1, no. 3 (1978): 215–239.

4. Richard J. Trudeau, *Introduction to Graph Theory* (New York: Dover Books, 2001).

5. Maarten van Steen, *Graph Theory and Complex Networks: An Introduction* (author, 2010).

6. Tanguy Godquin et al., "Applied Graph Theory to Security: A Qualitative Placement of Security Solutions within IOT Networks," *Journal of Information Security and Applications* 55 (2020).

Chapter 4

1. Kanniga Devi Rangaswamy and Murugaboopathi Gurusamy, "Application of Graph Theory Concepts in Computer Networks and Its Suitability for the Resource Provisioning Issues in Cloud Computing—A Review," *Journal of Computer Science* 14, no. 2 (2018): 163–172.

2. Chris Sanders, *Practical Packet Analysis*, 3rd ed. (San Francisco: No Starch Press, 2017).

3. James Forshaw, *Attacking Network Protocols* (San Francisco: No Starch Press, 2017).

Chapter 5

1. Jon M. Kleinberg, "Authoritative Sources in a Hyperlinked Environment," *Journal of the ACM* 46, no. 5 (1999): 604–632.

2. Onook Oh, Manish Agrawal, and H. Raghav Rao, "Information Control and Terrorism: Tracking the Mumbai Terrorist Attack Through Twitter," *Information Systems Frontiers* 13, no. 1 (2011): 33–43.

3. Soumajyoti Sarkar et al., "Predicting Enterprise Cyber Incidents Using Social Network Analysis on Dark Web Hacker Forums," *Cyber Defense Review* (2019): 87–102, *https://www.jstor.org/stable/26846122.*

4. Sean Everton, "Tracking, Destabilizing and Disrupting Dark Networks with Social Networks Analysis," Dark Networks Course Manual, Naval Postgraduate School (2008).

Chapter 6

1. International Association of Crime Analysts (IACA), *Social Network Analysis for Law Enforcement*, ed. Daniel S. Polans (Overland Park, KS: IACA, 2018).

2. Jia Wei and Sebastian Schuetz, "A Social Network Analysis Perspective on Users' Vulnerability to Socially Engineered Phishing Attacks," in the *International Conference on Information Systems (ICIS) 2019 Proceedings* (Munich: ICIS, 2019), 37.

3. Agata Fronczak and Piotr Fronczak, "Biased Random Walks in Complex Networks: The Role of Local Navigation Rules," *Physical Review E* 80, no. 1 (2009): 016107.

4. Hootan Nakhost and Martin Müller, "Monte-Carlo Exploration for Deterministic Planning," in *Proceedings of the 21st International Joint Conference on Artificial Intelligence (IJCAI)*, ed. Hiroaki Kitano (San Francisco: Morgan Kaufmann, 2009), 1766–1771.

5. Afraa Attiah, Mainak Chatterjee, and Cliff C. Zou, "A Game Theoretic Approach to Model Cyber Attack and Defense Strategies," in *Proceedings of the 2018 IEEE International Conference on Communications* (Piscataway, NJ: IEEE, 2018), 1–7.

6. Tao Zhou, Linyuan Lü, and Yi-Cheng Zhang, "Predicting Missing Links via Local Information," *European Physical Journal B* 71, no. 4 (2009): 623–630.

7. George Bernard Dantzig and Delbert Ray Fulkerson, *On the Max Flow Min Cut Theorem of Networks* (Santa Monica, CA: RAND Corp., 1955).

8. Seth Godin, "Linchpin: Are You Indispensable?", *Teacher Librarian* 37, no. 4 (2010): 77.

9. Dantzig and Fulkerson, *Max Flow Min Cut.*

10. David Liben-Nowell and Jon Kleinberg, "The Link Prediction Problem for Social Networks," in *Proceedings of the Twelfth International Conference on Information and Knowledge Management* (New York: ACM, 2003), 556–559.

11. Wayne W. Zachary, "An Information Flow Model for Conflict and Fission in Small Groups," *Journal of Anthropological Research* 33, no. 4 (1977): 452–473.

Chapter 7

1. Mike Burmester, Ronald R. Rivest, and Adi Shamir, "Geometric Cryptography: Identification by Angle Trisection (Draft 2)" (unpublished manuscript, November 4, 1997), *https://people.csail.mit.edu/rivest/pubs/BRS97.pdf.*

Chapter 8

1. Nikolay Elenkov, *Android Security Internals* (San Francisco: No Starch Press, 2015).

2. Joshua J. Drake et al., *Android Hacker's Handbook* (Indianapolis, IN: John Wiley & Sons, 2014).

3. Charlie Miller et al., *iOS Hacker's Handbook* (Indianapolis, IN: John Wiley & Sons, 2014).

4. Linnet Taylor, "No Place to Hide? The Ethics and Analytics of Tracking Mobility Using Mobile Phone Data," *Environment and Planning D: Society and Space* 34, no. 2 (2016): 319–336.

5. Taylor, "No Place."

6. Shadowtalker, "Creating a Circle with Radius in Metres," Stack Exchange, August 31, 2022, *https://gis.stackexchange.com/questions/121256/creating-a-circle-with-radius-in-metres.*

7. Muneendra Kumar, "World Geodetic System 1984: A Modern and Accurate Global Reference Frame," *Marine Geodesy* 12, no. 2 (1988): 117–126.

8. Eric H., "Python: Shapely, Cascaded Intersections Within One Polygon," Stack Overflow, January 29, 2019, *https://stackoverflow.com/questions/54424392/python-shapely-cascaded-intersections-within-one-polygon.*

9. Eric H., "Python."

10. Lisa Vaas, "War Kitteh," Naked Security, August 12, 2014, *https://nakedsecurity.sophos.com/tag/war-kitteh.*

Chapter 9

1. Judith Summers, "Broad Street Pump Outbreak," accessed December 4, 2022, *https://www.ph.ucla.edu/epi/snow/broadstreetpump.html.*

2. A. Verma et al., "Rationalizing Police Patrol Beats Using Voronoi Tessellations," in *Proceedings of the 2010 IEEE International Conference on Intelligence and Security Informatics*, ed. C. Yang (Vancouver, BC: IEEE, 2010), 165–167.

Chapter 10

1. Niraj Chokshi, "Facial Recognition's Many Controversies, from Stadium Surveillance to Racist Software," *New York Times*, May 15, 2019, *https://www.nytimes.com/2019/05/15/business/facial-recognition-software-controversy.html.*

2. University of Essex, "Description of the Collection of Facial Images," n.d., *https://web.archive.org/web/20200810043556/https://cswww.essex.ac.uk/mv/allfaces/index.html.*

3. Left and right photos in Figures 10-3 and 10-4 are courtesy of Hassan Khan via Unsplash (*https://unsplash.com/photos/EGVccebWodM*) and Reza Biazar via Unsplash (*https://unsplash.com/photos/eSjmZW97cH8*), respectively.

4. The Facial Identification Scientific Working Group, "Facial Image Comparison Feature List for Morphological Analysis," accessed December 4, 2022, *https://fiswg.org/FISWG_Morph_Analysis_Feature_List_v2.0_20180911.pdf.*

5. scikit-learn, "Sklearn.feature_selection.mutual_info_classif," accessed December 4, 2022, *https://scikit-learn.org/stable/modules/generated/sklearn.feature_selection.mutual_info_classif.html.*

6. William Crumpler, "How Accurate Are Facial Recognition Systems—and Why Does It Matter?", Center for Strategic and International Studies, April 14, 2020, *https://www.csis.org/blogs/technology-policy-blog/how-accurate-are-facial-recognition-systems-%E2%80%93-and-why-does-it-matter.*

7. Aaron Holmes, "All It Takes to Fool Facial Recognition at Airports and Border Crossings Is a Printed Mask, Researchers Found," *Business Insider*, February 7, 2020, *https://www.businessinsider.com/facial-recognition-fooled-with-mask-kneron-tests-2019-12.*

Chapter 11

1. V. Chvátal, "A Combinatorial Theorem in Plane Geometry," *Journal of Combinatorial Theory, Series B* 18, no. 1 (1975): 39–41.
2. Mikael Pålsson and Joachim Ståhl, *The Camera Placement Problem: An Art Gallery Problem Variation* (master's thesis, Department of Computer Science, Lund University, 2008).
3. Jonathan Richard Shewchuk, "Delaunay Refinement Algorithms for Triangular Mesh Generation," *Computational Geometry* 22, no. 1–3 (2002): 21–74.
4. James Roland, "How Far Can We See and Why?", Healthline Media, May 23, 2019, *https://www.healthline.com/health/how-far-can-the-human-eye-see.*
5. Daniel Brélaz, "New Methods to Color the Vertices of a Graph," *Communications of the ACM* 22, no. 4 (1979): 251–256.
6. György Csizmadia and Géza Tóth, "Note on an Art Gallery Problem," *Computational Geometry* 10, no. 1 (1998): 47–55.

Chapter 12

1. Julia Courtenay, "Filming with Green Screen: Everything You Need to Know," InFocus Film School, June 10, 2021, *https://infocusfilmschool.com/filming-green-screen-guide.*
2. Jonathan Lazar, Jinjuan Heidi Feng, and Harry Hochheiser, *Research Methods in Human–Computer Interaction* (West Sussex, UK: Wiley, 2014).

Chapter 13

1. Janakiram MSV, "Tutorial: Blue/Green Deployments with Kubernetes and Istio," The New Stack, July 6, 2022, *https://thenewstack.io/tutorial-blue-green-deployments-with-kubernetes-and-istio.*
2. Nataliia Peterheria, "How to Form a Successful Development Team," Django Stars, December 26, 2022, *https://djangostars.com/blog/form-successful-development-team.*

INDEX

Page numbers referring to figures and tables are followed by an italicized *f* or *t*, respectively.

D

E

F

G

H

I

J

K

L

M

N

O

P

Q

R

T

X

Y

Z